REA's Books Are The Best...

They have rescued lots of grades and more!

(a sample of the hundreds of letters REA receives each year)

"Your books are great! They are very helpful, and have upped my grade in every class. Thank you for such a great product."

Student, Seattle, WA

"Your book has really helped me sharpen my skills and improve my weak areas. Definitely will buy more."

Student, Buffalo, NY

"Compared to the other books that my fellow students had, your book was the most useful in helping me get a great score."

Student, North Hollywood, CA

"I really appreciate the help from your excellent book. Please keep up your great work."

Student, Albuquerque, NM

"Your book was such a better value and was so much more complete than anything your competition has produced (and I have them all)!"

Teacher, Virginia Beach, VA

(more on next page)

(continued from previous page)

"Your books have saved my GPA, and quite possibly my sanity. My course grade is now an 'A', and I couldn't be happier."

Student, Winchester, IN

"These books are the best review books on the market. They are fantastic!"

Student, New Orleans, LA

"Your book was responsible for my success on the exam. . . I will look for REA the next time I need help."

Student, Chesterfield, MO

"I think it is the greatest study guide I have ever used!"

Student, Anchorage, AK

"I encourage others to buy REA because of their superiority. Please continue to produce the best quality books on the market."

Student, San Jose, CA

"Just a short note to say thanks for the great support your book gave me in helping me pass the test . . . I'm on my way to a B.S. degree because of you !"

Student, Orlando, FL

ALGEBRA & TRIGONOMETRY

By the Staff of
Research & Education Association
Dr. M. Fogiel, Director

Research & Education Association
61 Ethel Road West
Piscataway, New Jersey 08854

SUPER REVIEW®
OF ALGEBRA & TRIGONOMETRY

Year 2002 Printing

Printed in the United States of America

Library of Congress Control Number 00-130286

International Standard Book Number 0-87891-177-4

I-4

WHAT THIS Super Review WILL DO FOR YOU

This Super Review provides all that you need to know to do your homework effectively and succeed on exams and quizzes.

The book focuses on the core aspects of the subject, and helps you to grasp the important elements quickly and easily.

Outstanding Super Review features:

- Topics are covered in logical sequence
- Topics are reviewed in a concise and comprehensive manner
- The material is presented in student-friendly language that makes it easy to follow and understand
- Individual topics can be easily located
- Provides excellent preparation for midterms, finals and in-between quizzes
- In every chapter, reviews of individual topics are accompanied by Questions Q and Answers A that show how to work out specific problems
- At the end of most chapters, quizzes with answers are included to enable you to practice and test yourself to pinpoint your strengths and weaknesses
- Written by professionals and test experts who function as your very own tutors

Dr. Max Fogiel
Program Director

CONTENTS

1.1 Sets

A set is defined as a collection of items. Each individual item belonging to a set is called an element or member of that set. Sets are usually represented by capital letters, elements by lowercase letters. If an item k belongs to a set A, we write $k \in A$ ("k is an element of A"). If k is not in A, we write $k \notin A$ ("k is not an element of A"). The order of the elements in a set does not matter:

$$\{1, 2, 3\} = \{3, 2, 1\} = \{1, 3, 2\}, \text{etc.}$$

A set can be described in two ways: 1) it can be listed element by element, or 2) a rule characterizing the elements in a set can be formulated. For example, given the set A of the whole numbers starting with 1 and ending with 9, we can describe it either as $A = \{1, 2, 3, 4, 5, 6, 7, 8, 9\}$ or as {the set of whole numbers greater than 0 and less than 10}. In both methods, the description is enclosed in brackets. A kind of shorthand is often used for the second method of set description; instead of writing out a complete sentence in between the brackets, we write instead

$$A = \{k \mid 0 < k < 10, k \text{ is a whole number}\}$$

This is read as "the set of all elements k such that k is greater than 0 and less than 10, where k is a whole number."

A set not containing any members is called the empty or null set. It is written either as ϕ or { }.

Problem Solving Examples:

List the elements of the set:

$$\{n \mid 27 \leq n \leq 216, n \text{ is a perfect cube}\}$$

This expression may be read as: the set of all elements (numbers) n, such that n is a perfect cube between 27 and 216 inclusive. One way to list the elements is:

$$C = \{27, 64, 125, 216\}$$

Note that we could have chosen any other letter for the set, but "C" suggests the word "cube."

Given that $N = \{9, 15, 21, \ldots, 99\}$, describe N in words.

First, we observe that 9, 15, 21, and 99 are all odd numbers. However, two odd numbers are missing between 9 and 15 and also between 15 and 21. If we continue skipping two odd numbers, we should finally arrive at 99. So we can describe the set as: all third odd numbers between 9 and 99, inclusive. (Though technically correct, this is an awkward construction.)

We notice also that 9, 15, 21, and 99 are all odd multiples of 3. So an alternative description would be: the set of odd multiples of 3 between 9 and 99 inclusive. (This is a more elegant description.)

Q List the elements of the set:
{p | p is a quadrilateral whose diagonals bisect each other at right angles}

A In words, this may be read as the set of all elements, p, such that p is a quadrilateral whose diagonals bisect each other at right angles. There are only two such quadrilaterals, namely, the square and the rhombus.

Note that "rectangle" and "parallelogram" are not acceptable responses, since the diagonals of all rectangles do not necessarily bisect each other at right angles and the diagonals of all parallelograms do not necessarily bisect each other at right angles.

1.2 Subsets

Given two sets A and B, A is said to be a subset of B if every member of set A is also a member of set B. A is a *proper* subset of B if B contains at least one element not in A. We write $A \subseteq B$ if A is a subset of B, and $A \subset B$ if A is a proper subset of B.

Two sets are equal if they have exactly the same elements; in addition, if $A = B$ then $A \subseteq B$ and $B \subseteq A$.

Example: Let

$$A = \{1, 2, 3, 4, 5\}$$

$$B = \{1, 2\}$$

$$C = \{1, 4, 2, 3, 5\}$$

Then 1) A equals C, and A and C are subsets of each other, but not proper subsets, and 2) $B \subseteq A$, $B \subseteq C$, $B \subset A$, $B \subset C$. (B is a subset of both A and C. In particular, B is a proper subset of A and C.)

A universal set U is a set from which other sets draw their members. If A is a subset of U, then the complement of A, denoted A', is the set of all elements in the universal set that are not elements of A.

Example: If $U = \{1, 2, 3, 4, 5, 6, \ldots\}$ and $A = \{1, 2, 3\}$, then $A' = \{4, 5, 6, \ldots\}$.

Figure 1.1 illustrates this concept through the use of a Venn diagram.

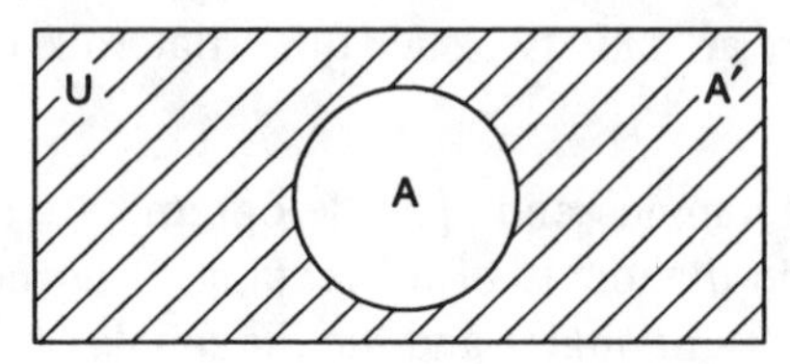

Figure 1.1

Problem Solving Examples:

List all the subsets of $C = \{1, 2\}$.

$\{1\}, \{2\}, \{1, 2\}$, and ϕ where ϕ is the empty set. Each set listed in the solution contains at least one element of the set C. The set $\{2, 1\}$ is identical to $\{1, 2\}$ and therefore is not listed. ϕ is included in the solution because ϕ is a subset of every set.

Find four proper subsets of $P = \{n \mid n \in I, -5 < n \leq 5\}$. ($I$ stands for integer.)

$P = \{-4, -3, -2, -1, 0, 1, 2, 3, 4, 5\}$. All these elements are integers that are less than or equal to 5 and greater than -5. A set A is a proper subset of P if every element of A is an element of P and in addition there is an element of P that is not in A.

A) $B = \{-4, -2, 0, 2, 4\}$ is a subset because each element of B is an integer greater than -5 but less than or equal to 5. B is a proper subset because 3 is an element of P but not an element of B. We can write $3 \in P$ but $3 \notin B$.

B) $C = \{3\}$ is a subset of P, since $3 \in P$. However, $5 \in P$ but $5 \notin C$. Hence, $C \subset P$.

C) $D = \{-4, -3, -2, -1, 1, 2, 3, 4, 5\}$ is a proper subset of P, since each element of D is an element of P, but $0 \in P$ and $0 \notin D$.

D) $\phi \subset P$, since ϕ has no elements. Note that ϕ is the empty set. ϕ

is a proper subset of every set except itself. Note that there are many other possible sets which comprise the proper subset of P.

1.3 Union and Intersection of Sets

The union of two sets A and B, denoted $A \cup B$, is the set of all elements that are either in A or B or both.

The intersection of two sets A and B, denoted $A \cap B$ is the set of all elements that belong to both A and B.

If $A = \{1, 2, 3, 4, 5\}$ and $B = \{2, 3, 4, 5, 6\}$, then $A \cup B = \{1, 2, 3, 4, 5, 6\}$ and $A \cap B = \{2, 3, 4, 5\}$.

If $A \cap B = \phi$, then A and B have no elements in common and are said to be *disjointed.* Figures 1.2 and 1.3 are Venn diagrams for union and intersection. The shaded areas represent the given operation.

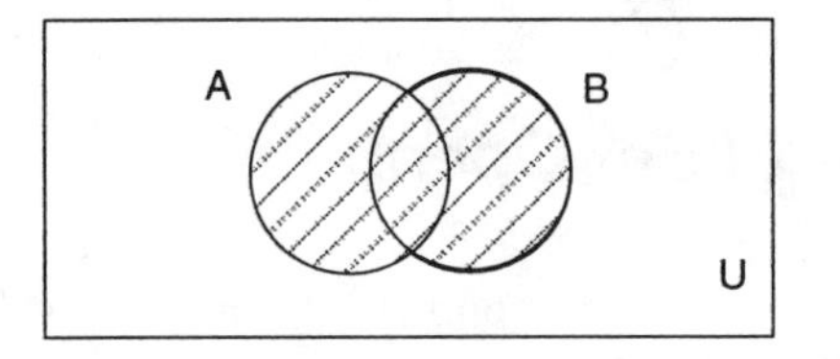

A U B

Figure 1.2

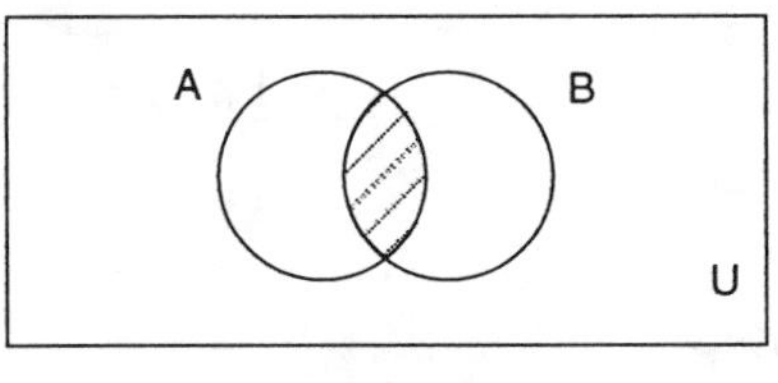

A ∩ B

Figure 1.3

Problem Solving Examples:

If $A = \{1, 2, 3, 4, 5\}$ and $B = \{2, 3, 4, 5, 6\}$, find $A \cup B$.

The symbol $\cup$ is used to denote the union of sets. Thus, $A \cup B$ (which is read "the union of A and B) is the set of all elements that are in either A or B or both. In this problem, if

$$A = \{1, 2, 3, 4, 5\} \text{ and } B = \{2, 3, 4, 5, 6\},$$

then

$$A \cup B = \{1, 2, 3, 4, 5, 6\}.$$

If $A = \{1, 2, 3, 4, 5\}$ and $B = \{2, 3, 4, 5, 6\}$, find $A \cap B$.

The intersection of two sets A and B is the set of all elements that belong to both A and B; that is, all elements common to A and B. In this problem, if

$$A = \{1, 2, 3, 4, 5\} \text{ and } B = \{2, 3, 4, 5, 6\},$$

then

$$A \cap B = \{2, 3, 4, 5\}.$$

1.4 Laws of Set Operations

If U is the universal set and A, B, and C are *any* subsets of U, then the following hold for union, intersection, and complement:

Identity Laws

1a. $A \cup \phi = A$ 1b. $A \cap \phi = \phi$

2a. $A \cup U = U$ 2b. $A \cap U = A$

Idempotent Laws

3a. $A \cup A = A$ 3b. $A \cap A = A$

Complement Laws

4a. $A \cup A' = U$ 4b. $A \cap A' = \phi$

5a. $(A')' = A$ 5b. $\phi' = U$; $U' = \phi$

Commutative Laws

6a. $A \cup B = B \cup A$ 6b. $A \cap B = B \cap A$

Associative Laws

7a. $(A \cup B) \cup C = A \cup (B \cup C)$

7b. $(A \cap B) \cap C = A \cap (B \cap C)$

Distributive Laws

8a. $A \cup (B \cap C) = (A \cup B) \cap (A \cup C)$

8b. $A \cap (B \cup C) = (A \cap B) \cup (A \cap C)$

De Morgan's Laws

9a. $(A \cup B)' = A' \cap B'$ 9b. $(A \cap B)' = A' \cup B'$

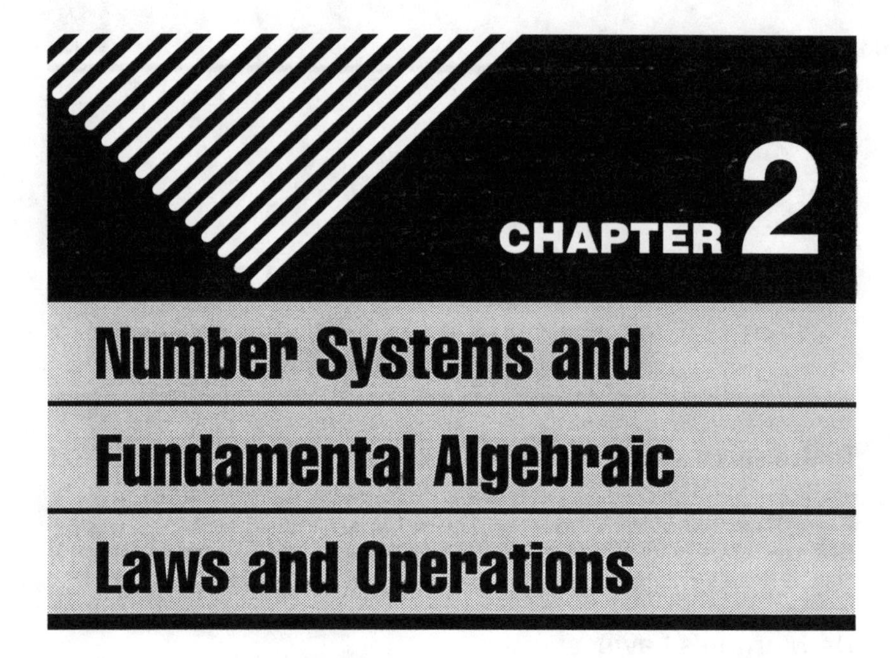

2.1 Decimals

If we divide the denominator of a fraction into its numerator, we obtain a decimal form for it. This form attaches significance to the placement of an integer relative to a decimal point. The first place to the left of the decimal point is the units place; the second to the left is the tens; third, the hundreds, etc. The first place to the right of the decimal point is the tenths, the second the hundredths, etc. The integer in each place tells how many of the values of that place the given number has.

Example: 721 has seven hundreds, two tens, and one unit.
.584 has five tenths, eight hundredths, and four thousandths.

Since a rational number is of the form $\frac{a}{b}$, $b \neq 0$, then all rational numbers can be expressed as decimals by dividing b into a. The resulting decimal is either a terminating decimal, meaning that b divides a with remainder 0 after a certain point; or repeating, meaning that b

continues to divide a so that the decimal has a repeating pattern of integers.

Example:

A) $\frac{1}{2} = .5$

B) $\frac{1}{3} = .3\overline{3}$

C) $\frac{11}{16} = .6875$

D) $\frac{2}{7} = .285714\overline{285714}$

A) and C) are terminating decimals; B) and D) are repeating decimals. This explanation allows us to define irrational numbers as numbers whose decimal form is nonterminating and nonrepeating.

Example: $\sqrt{3} = 1.732\ldots$

So the set of reals is the union of the set of rationals and the set of irrationals ($\mathbb{R} = \mathbb{Q} \cup \mathbb{Q}'$).

Problem Solving Examples:

Write $\frac{2}{7}$ as a repeating decimal.

To write a fraction as a repeating decimal, divide the numerator by the denominator, until a pattern of repeated digits appears:

$$2 \div 7 = .285714285714\ldots$$

Identify the entire portion of the decimal which is repeated. The repeating decimal can then be written in the shortened form:

$$\frac{2}{7} = \overline{.285714}$$

Find the common fraction form of the repeating decimal 0.4242....

Let x represent the repeating decimal.

$$x = 0.4242\ldots$$

$$100x = 42.42\ldots \qquad \text{multiplying by 100}$$

$$-\quad x = 0.42\ldots \qquad \text{subtracting } x \text{ from } 100x$$

$$99x = 42$$

Divide both sides of the difference by 99, and reduce the fraction.

$$\frac{99x}{99} = \frac{42}{99}$$

$$x = \frac{42}{99} = \frac{14}{33}$$

The repeating decimal of this example had two digits that repeated. The first step in the solution was to multiply both sides of the original equation by the 2nd power of 10, or 10^2 or 100. If there were three digits that repeated, the first step in the solution would be to multiply both sides of the original equation by the 3rd power of 10, or 10^3 or 1,000.

Find $0.25\overline{25}$ as a quotient of integers.

Let $x = 0.25\overline{25}$. Multiply both sides of this equation by 100:

$$100x = 100\,(0.25\overline{25}) \qquad (1)$$

Multiplying by 100 is equivalent to moving the decimal point two places to the right, and since the digits 25 are repeated, we have:

$$100x = 25.25\overline{25} \qquad (2)$$

Now subtract Equation (1) from Equation (2):

$$\begin{aligned} 100x &= 25.25\overline{25} \\ -\ x &= 0.25\overline{25} \\ \hline 99x &= 25.0000 \end{aligned}$$

or $$99x = 25 \qquad (3)$$

(Note that this operation eliminates the decimal point.)

Dividing both sides of Equation (3) by 99:

$$\frac{99x}{99} = \frac{25}{99}$$

$$x = \frac{25}{99}$$

Therefore,

$$0.2525 = x = \frac{25}{99}$$

Also, note that the given repeating decimal, $0.25\overline{25}$, was multiplied by 100 or 10^2, where the power of 10 (which is 2) is the same as the number of repeating digits (namely, 2) for this problem. In general, for problems of this type, if the repeating decimal has n repeating digits, then the repeating decimal should be multiplied by 10^n.

2.2 Number Systems

Most of the numbers used in algebra belong to a set called the real numbers, or reals. This set, denoted as $\mathbb{R}$, can be represented graphically by the real number line (Figure 2.1).

Given a straight horizontal line extending continuously in both directions, we arbitrarily fix a point and label it with the number 0. In a similar manner, we can label any point on the line with one of the real numbers, depending on its position relative to 0. Numbers to the right of zero are called positive, while those to the left are called nega-

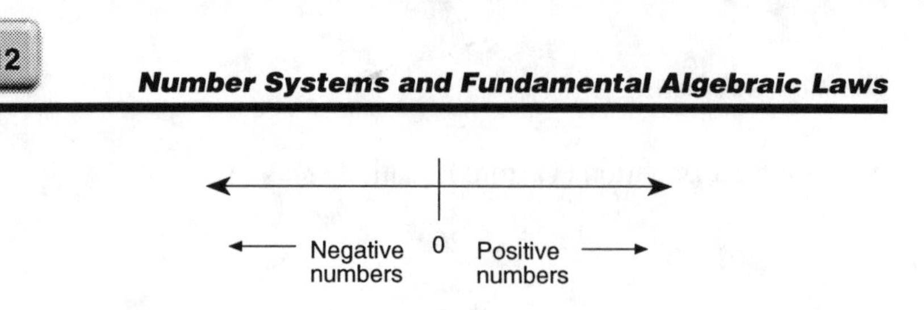

Figure 2.1

tive. Value increases from left to right, so that if a is to the right of b, it is said to be greater than b.

2.2.1 Integers

If we divide the number line into equal segments called unit lengths, we can then label the boundary points of these segments according to their distance from zero. For example, the point two lengths to the left of zero is –2, while the point three lengths to the right of 0 is +3 (the + sign is usually assumed, so +3 is written as 3). The number line now looks like the one in Figure 2.2.

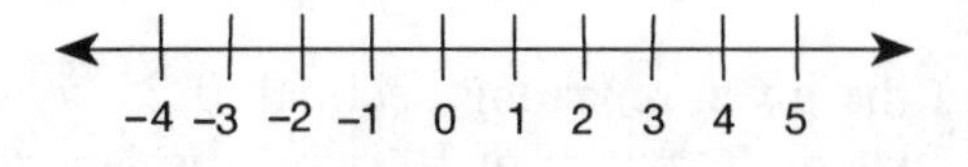

Figure 2.2

These boundary points represent the subset of the reals known as the integers, denoted $\mathbb{Z}$. Some subsets of $\mathbb{Z}$ are the natural numbers or positive integers, the set of integers starting with 1 and increasing, $\mathbb{Z}+ = \mathbb{N} = \{1, 2, 3, 4,\ldots\}$; the whole numbers, the set of integers starting with 0 and increasing, $\mathbb{W} = \{0, 1, 2, 3,\ldots\}$; the negative integers, the set of integers starting with –1 and decreasing: $\mathbb{Z}- = \{-1, -2, -3,\ldots\}$; and the prime numbers, the set of positive integers greater than 1 that are divisible only by 1 and themselves: $\{2, 3, 5, 7, 11,\ldots\}$.

2.2.2 Rationals

One of the main subsets of the reals is the set of rational numbers, denoted $\mathbb{Q}$. This set is defined as all the numbers that can be expressed in the form a/b, where a and b are integers, $b \neq 0$, or as a decimal that terminates or an infinitely repeating decimal. This form is called a fraction or ratio; *a* is known as the numerator, or as a decimal that terminates or an infinitely repeating decimal, *b* the denominator.

Example: $\frac{-7}{5}, \frac{8}{6}, \frac{9}{-3}, \frac{5}{100}$

Note: The integers can all be expressed in the form *a*/*b*.

Example: $2 = \frac{2}{1}, \quad -3 = \frac{6}{-2}, \quad -5 = \frac{-5}{1}$

2.2.3 Irrationals

The complement of the set of rationals is the irrationals, whose symbol is $\mathbb{Q}'$. For now, they are defined as the set of real numbers that cannot be expressed in the form

$$\frac{a}{b}, b \neq 0.$$

Problem Solving Examples:

Which of the following are *rational* quantities?

A) 5π

B) $\frac{81}{25}$

C) 7.25

D) $\sqrt{\frac{36}{144}}$

E) $\sqrt[3]{\frac{27}{49}}$

We can verify whether any of these quantities are rational by first re-writing them:

A) Since π is an irrational quantity (infinite, non-repeating), any multiple of it with n ($n \neq 0$) is still an *irrational* quantity.

B) $\frac{81}{25} = 3.24$ is a rational quantity. So $\frac{81}{25}$ is a *rational.*

C) 7.25 is clearly a *rational* quantity.

D) $\sqrt{\frac{36}{144}} = \frac{6}{12} = 0.5$, which is a *rational* quantity.

E) $\sqrt[3]{\frac{27}{49}} = \frac{3}{\sqrt[3]{49}}$. Since $\sqrt[3]{49}$ is *not* rational, $\frac{3}{\sqrt[3]{49}}$ is not rational and neither is $\sqrt[3]{\frac{27}{49}}$.

Which of the following are *irrational* quantities?

A) $\frac{\pi}{5\pi}$

B) $\sqrt[3]{125}$

We can verify whether any of these quantities is irrational by first rewriting it:

A) $\frac{\pi}{5\pi} = \frac{1}{5} = 0.2$, which is *not* an irrational quantity.

B) $\sqrt[3]{125} = 5$, which is *not* an irrational quantity.

Given $Q = \frac{n}{d}$, $n = 2, 3, 4, \ldots$, list the first four integers that substitute for d, greater than 1, which make Q a rational number.

We are looking for the first four integers greater than 1 which, when used as denominators with 2, 3, 4, ... as numerators, yield either an integer or a terminating decimal. The required integers are 2, 3, 4, and 5.

Check this by dividing any integer by 2, 4, and 5. [In general, when any integer is divided by 2, 4, 5, or their powers (4, 8, 16, …; 16, 64, 256, …; 25, 125, 625, …, respectively) or products (2×4, 2×5, 4×5, etc.), the quotient is always a rational number.]

2.3 Absolute Value

The absolute value of a real number A is defined as follows:

$$|A| = \begin{cases} A & \text{if } A \geq 0 \\ -A & \text{if } A < 0 \end{cases}$$

Example: $|5| = 5$, $|-8| = -(-8) = 8$.

Absolute values follow the given rules:

A) $|-A| = |A|$

B) $|A| \geq 0$, equality holding only if $A = 0$

C) $\left|\frac{A}{B}\right| = \frac{|A|}{|B|}$, $B \neq 0$

D) $|AB| = |A| \times |B|$

E) $|A|^2 = A^2$

Absolute value can also be expressed on the real number line as the distance of the point represented by the real number from the point labeled 0.

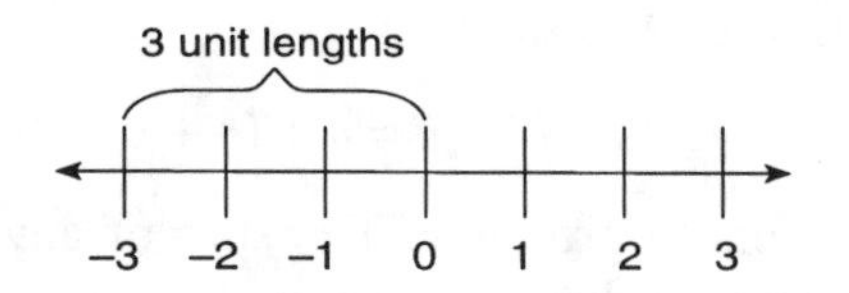

So $|-3| = 3$ because -3 is 3 units to the left of 0.

2.4 Fundamental Algebraic Laws

Note that $a, b, c \in \mathbb{R}$.

A) Closure Law of Addition:

$$a + b = c.$$

The sum of two real numbers is always a real number.

B) Closure Law of Multiplication:

$$a \times b = c.$$

The product of two real numbers is always a real number.

C) Commutative Law of Addition:

$$a + b = b + a.$$

Commutative refers to position. The sum of two real numbers is the same even if their positions are changed.

Example: $3 + 2 = 5 = 2 + 3.$

D) Commutative Law of Multiplication:

$$a \times b = b \times a.$$

The product of two real numbers is the same even if their positions are changed.

Example: $3 \times 2 = 6 = 2 \times 3.$

E) Associative Law of Addition:

$$(a + b) + c = a + (b + c).$$

Associative refers to grouping. The sum of any three real numbers is the same regardless of the way they are grouped.

Example: $(5 + 3) + 2 = 10 = 5 + (3 + 2).$

F) Associative Law of Multiplication:

$$(a \times b) \times c = a \times (b \times c).$$

The product of any three real numbers is the same, regardless of the way they are grouped.

Example: $(5 \times 3) \times 2 = 30 = 5 \times (3 \times 2)$.

G) Additive Identity:

There exists a real number 0 such that $a + 0 = a$. The number 0 is referred to as the additive identity.

H) Multiplicative Identity:

There exists a real number 1 such that $a \times 1 = a$. The number 1 is referred to as the multiplicative identity.

I) Additive Inverse:

For each real number a, there is a unique real number $-a$, called the additive inverse of a, such that $a + (-a) = 0$.

Example: $7 + (-7) = 0$.

J) Multiplicative Inverse:

For every number a, $a \neq 0$, there is a unique real number $\frac{1}{a}$ called the multiplicative inverse of a, such that $a \times \frac{1}{a} = 1$.

Example: $7 \times \frac{1}{7} = 1$.

K) Zero Law:

For every number a, $a \times 0 = 0$.

L) Distributive Law for Multiplication with Respect to Addition and Subtraction:

$$\begin{aligned} a(b + c) &= ab + ac \\ &= ba + ca \quad \text{by the commutative law} \\ &= (b + c)\,a. \end{aligned}$$

Also $a(b - c) = ab - ac$

$= ba - ca$

$= (b - c)a.$

Example:

1) $3(4 + 5) = 3\ (4) + 3(5)$
$= (4)3 + (5)3 = (4 + 5)3 = 27$

2) $3(5 - 4) = 3(5) - 3(4)$
$= (5)3 - (4)3 = (5 - 4)3 = 3.$

These rules also hold for certain subsets of the reals $\mathbb{R}$, such as the rationals $\mathbb{Q}$. They do not hold for all subsets of $\mathbb{R}$, however; for instance, the integers $\mathbb{Z}$ do not contain multiplicative inverses for integers other than 1 or –1.

Problem Solving Examples:

Find the sum $8 + (-3)$.

The sum of $8 + (-3)$ can be obtained by using facts from arithmetic and the associative law:

$$8 + (-3) = (5 + 3) + (-3)$$

Using the associative law of addition $(a + b) + c = a + (b + c)$:

$$= 5 + [3 + (-3)]$$

Using the additive inverse property, $a + (-a) = 0$:

$$= 5 + 0$$

Using the additive identity property, $a + 0 = a$:

$$= 5$$

Q Show that $(-2) + (-3) = -5$.

This small problem illustrates some of the basic ideas involved in mathematical proof. We know that $(-2) + (-3)$ is an integer because the integers are closed under addition. To show that this integer is –5, we ask ourselves what property is characteristic of –5. Thus,

$$5 + (-5) = 0$$

by the additive inverse property, $a + (-a) = 0$. Moreover, –5 is the only number which when added to 5 gives 0; for if $(5 + b) = 0$, by the additive identity, $a = a + 0$, $-5 = -5 + 0 = -5 + (5 + b)$, by our hypothesis, $5 + b = 0$

$= (-5 + 5) + b$	by associative law of addition $a + (b + c) = (a + b) + c$;
$= 0 + b$	by additive inverse property, $a + (-a) = 0$;
$= b + 0$	by commutative law of addition, $a + b = b + a$;
$= b$	by additive identity, $a + 0 = a$.

Thus, $-5 = b$, proving that –5 is the only number which when added to 5 gives 0.

We therefore see that $(-2) + (-3) = -5$ if and only if $5 + [(-2) + (-3)] = 0$. We show below that this sum is zero.

$5 + [(-2) + (-3)] = [3 + 2] + [(-2) + (-3)]$

$= 3 + \{2 + [(-2) + (-3)]\}$	by the associative law of addition, $(a + b) + c = a + (b + c)$;
$= 3 + [0 + (-3)]$	by the additive inverse property, $a + (-a) = 0$;
$= 3 + [(-3) + 0]$	by the commutative law of addition, $a + b = b + a$;
$= 3 + (-3)$	by additive identity property, $a + 0 = a; = 0$;
$= 0$	by additive inverse property, $a + (-a) = 0$.

Thus, we have shown

A) $5 + (-5) = 0$

B) (–5) is the only number which when added to 5 equals 0

C) $5 + (-5) = 0 = 5 + [(-2) + (-3)]$ and, therefore, $(-5) = (-2) + (-3)$, completing our proof.

2.5 Basic Algebraic Operations

A) To add two numbers with like signs, add their absolute values and prefix the sum with the common sign.

Example: $6 + 2 = 8$, $(-6) + (-2) = -8$

B) To add two numbers with unlike signs, find the difference between their absolute values, and prefix the result with the sign of the number with the greater absolute value.

Example: $(-4) + 6 = 2$, $15 + (-19) = -4$

C) To subtract a number b from another number a, change the sign of b and add to a.

Examples:

(1) $10 - (3) = 10 + (-3) = 7$

(2) $2 - (-6) = 2 + 6 = 8$

(3) $(-5) - (-2) = -5 + (+2) = -3$

D) To multiply (or divide) two numbers having like signs, multiply (or divide) their absolute values and prefix the result with a positive sign.

Examples:

(1) $(5)(3) = 15$

(2) $\frac{-6}{-3} = 2$

E) To multiply (or divide) two numbers having unlike signs, multiply (or divide) their absolute values and prefix the result with a negative sign.

Examples:

(1) $(-2)(8) = -16$

(2) $\frac{9}{-3} = -3$

2.6 Operations with Fractions

To understand the operations on fractions, it is first desirable to understand what is known as factoring.

The product of two numbers is equal to a unique number. The two numbers are said to be factors of the unique number and the process of finding the two numbers is called factoring. It is important to note that when a number in a particular set is factored, then the factors of the number are also in the same set.

Example: the factors of 6 are

1) 1 and 6 since $1 \times 6 = 6$

2) 2 and 3 since $2 \times 3 = 6$

A) The value of a fraction remains unchanged if its numerator and denominator are both multiplied or divided by the same number, other than zero.

Example: $\frac{1}{2} \times \frac{2}{2} = \frac{2}{4} = \frac{1}{2}$

This is because a fraction $\frac{b}{b}$, where b is any number, is equal to the multiplicative identity 1.

B) To simplify a fraction is to convert it into a form in which the numerator and denominator have no common factor other than 1.

Example:
$$\frac{50}{25} = \frac{50 \div 25}{25 \div 25}$$
$$= \frac{2}{1} = 2$$

C) The algebraic sum of the fractions having a common denominator is a fraction whose numerator is the algebraic sum of the numerators of the given fractions and whose denominator is the common denominator.

Example: $\frac{11}{3} + \frac{5}{3} = \frac{11+5}{3} = \frac{16}{3}$

Similarly, for subtraction,

$$\frac{11}{3}-\frac{5}{3}=\frac{11-5}{3}=\frac{6}{3}=2$$

D) To find the sum of two fractions having different denominators, it is necessary to find the lowest common denominator (LCD) of the different denominators and convert the fractions into equivalent fractions having the lowest common denominator as a denominator.

Example: $$\frac{11}{6}+\frac{5}{16}=?$$

To find the LCD, we must first find the prime factors of the two denominators.

$$6=2\times 3$$

$$16=2\times 2\times 2\times 2$$

$$\text{LCD}=2\times 2\times 2\times 2\times 3=48$$

We do not need to repeat the 2 that appears as a factor of 6, since 2 also appears several times as a factor of 16.

We now rewrite 11/6, 5/16 to have 48 as their denominator.

$$\frac{11}{6}\times\frac{8}{8}=\frac{88}{48}$$

$$\frac{5}{16}\times\frac{3}{3}=\frac{15}{48}$$

We may now apply rule C) to find $\frac{11}{6}+\frac{5}{16}=\frac{88}{48}+\frac{15}{48}=\frac{103}{48}$.

E) The product of two or more fractions produces a fraction whose numerator is the product of the numerators of the given fractions and whose denominator is the product of the denominators of the given fractions.

Example: $$\frac{2}{3}\times\frac{1}{5}\times\frac{4}{7}=\frac{8}{105}$$

F) The quotient of two given fractions is obtained by inverting the divisor and then multiplying.

Example: $\frac{8}{9} \div \frac{1}{3} = \frac{8}{9} \times \frac{3}{1} = \frac{8}{3}$

2.7 Imaginary and Complex Numbers

2.7.1 Imaginary Numbers ($\mathbb{R}'$)

If a number a is multiplied by itself to produce a new number b, b is called the square of a and a the square root of b, denoted $a = \sqrt{b}$.

Example: $3 \times 3 = 3^2 = 9, \quad \sqrt{9} = 3$

3^2, read "3 to the second power" or "3 squared," indicates that 3 is to be used as a factor of the expression 3^2, i.e., $3^2 = 3 \times 3$.

According to the law of signs for real numbers, the square of a positive or negative number is always positive. This means that it is impossible to take the square root of a negative number in the real number system. In order to make this possible, the symbol i is defined as $i = \sqrt{-1}$, $i^2 = -1$. i is called an imaginary number, as is any multiple of i by a real number. This set is denoted $\mathbb{R}'$.

2.7.2 Complex Numbers ($\mathbb{C}$)

A complex number is a combination of real and imaginary numbers of the form $a + bi$, where a and b are real and i is defined as above. a is called the real part while bi is called the imaginary part of $a + bi$.

Both the real and the imaginary number sets are subsets of $\mathbb{C}$. $\mathbb{R} = \{a + bi \mid b = 0\}$ and $\mathbb{R}' = \{a + bi \mid a = 0\}$. Figure 2.3 uses a Venn diagram to illustrate the relationships of the various number systems to each other, while Figure 2.4 uses the tree form.

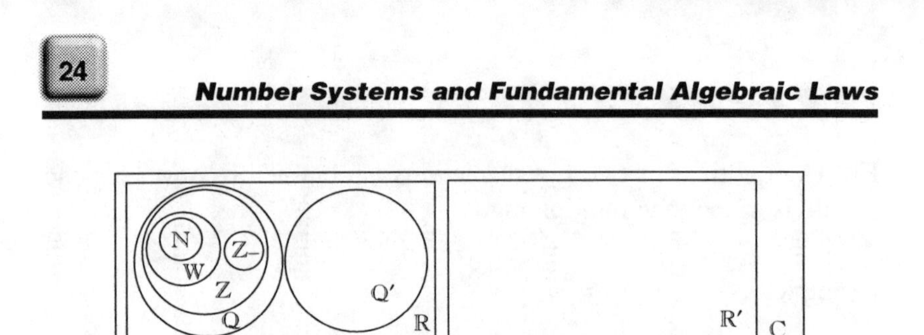

Figure 2.3

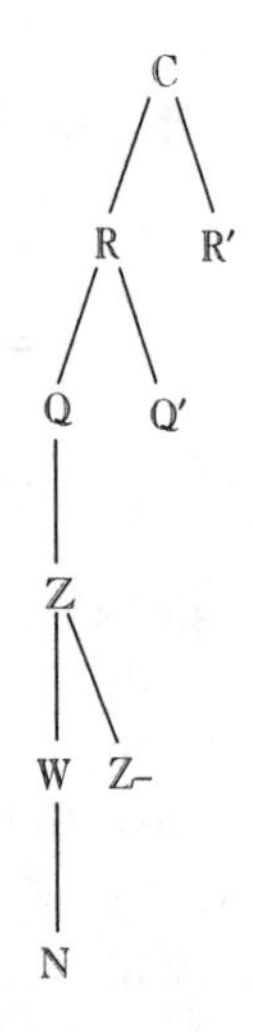

Figure 2.4

CHAPTER 3

Exponents and Radicals

3.1 Exponents

Given the expression $a^n = b$, where a, n, and $b \in$ R, a is called the base, n is called the exponent or power.

Example: In 3^2, 3 is the base, 2 is the exponent. If n is a positive integer, and if x and y are real numbers such that $x^n = y$, then x is said to be an nth root of y, written

$$x = \sqrt[n]{y} = y^{1/n}.$$

Positive Integral Exponent:

If n is a positive integer, then a^n represents the product of n factors, each of which is a.

Negative Integral Exponent:

If n is a positive integer,

$$a^{-n} = \frac{1}{a^n}, \quad a \neq 0$$

Example: $$2^{-4} = \frac{1}{2^4} = \frac{1}{2 \times 2 \times 2 \times 2} = \frac{1}{16}$$

Positive Fractional Exponent:

$$a^{m/n} = \sqrt[n]{a^m}$$

where m and n are positive integers.

Example: $4^{3/2} = \sqrt[2]{4^3} = \sqrt[2]{64} = 8$

or $4^{3/2} = (\sqrt[2]{4})^3 = 2^3 = 8$

Negative Fractional Exponent:

$$a^{-m/n} = \frac{1}{a^{m/n}}$$

Example: $27^{-2/3} = \frac{1}{27^{2/3}} = \frac{1}{\sqrt[3]{27^2}} = \frac{1}{\sqrt[3]{729}} = \frac{1}{9}$

or $27^{-2/3} = \frac{1}{27^{2/3}} = \frac{1}{(\sqrt[3]{27})^2} = \frac{1}{3^2} = \frac{1}{9}$

Zero Exponent: $a^0 = 1, \ a \neq 0$

General Laws of Exponents:

A) $a^p a^q = a^{p+q}$

B) $(a^p)^q = a^{pq}$

C) $\frac{a^p}{a^q} = a^{p-q}, \quad a \neq 0$

D) $(ab)^p = a^p b^p$

E) $\left(\frac{a}{b}\right)^p = \frac{a^p}{b^p}, \quad b \neq 0$

Problem Solving Examples:

Perform the indicated operations:
$(7 \times 10^5)^3 \times (3 \times 10^{-3})^4$.

Since $(ab)^x = a^x b^x$,
$(7 \times 10^5)^3 \times (3 \times 10^{-3})^4 = (7)^3(10^5)^3 \times (3)^4(10^{-3})^4$.

Recall that $(ab)^x = a^x b^x$ and that $(a^x)^y = a^{xy}$. Thus,

$$= (7^3)(10^{5 \times 3}) \times (3^4)(10^{-3 \times 4})$$

$$= (7^3)(10^{15}) \times (3^4)(10^{-12})$$

$$= (7^3)(3^4)(10^{15})(10^{-12}).$$

Since $a^x \times a^y = a^{x+y}$

$$= (7^3)(3^4)(10^{15+(-12)})$$

$$= 7^3 3^4 10^3 = (343)\ (81)\ (1{,}000)$$

$$= 27{,}783{,}000.$$

Simplify:

A) $2^3 \times 2^2$

B) $a^3 \times a^5$

C) $x^6 \times x^4$

If a is any number and n is any positive integer, the product of the n factors $a \times a \times a \ldots a$ is denoted by a^n. a is called the base and n is called the exponent. Also, for base a and exponents m and n, m and n being positive integers, we have the law:

$$a^m \times a^n = a^{m+n}.$$

Therefore,

A) $2^3 \times 2^2 = 2^{3+2} = 2^5 = 32$

or $2^3 \times 2^2 = (2 \times 2 \times 2)(2 \times 2) = 8 \times 4 = 32$

B) $a^3 \times a^5 = a^{3+5} = a^8$

or $a^3 \times a^5 = (a \times a \times a)(a \times a \times a \times a \times a)$

$= (a \times a \times a \times a \times a \times a \times a \times a) = a^8$

C) $x^6 \times x^4 = x^{6+4} = x^{10}$

or $(x \times x \times x \times x \times x \times x)\ (x \times x \times x \times x) = x^{10}$

Use the laws of exponents to perform the indicated operations:

A) $5x^5 \times 2x^2$

B) $(x^4)^6$

C) $\dfrac{8y^8}{2y^2}$

D) $\dfrac{x^3}{x^6}\left(\dfrac{7}{x}\right)^2$

Noting the following properties of exponents:

1) $a^b \times a^c = a^{b+c}$

2) $(a^b)^c = a^{b \times c}$

3) $\dfrac{a^b}{a^c} = a^{b-c}$

4) $\left(\dfrac{a}{b}\right)^c = \dfrac{a^c}{b^c}$

We proceed to evaluate these expressions.

A) $5x^5 \bullet 2x^2 = 5 \bullet 2 \bullet x^5 \bullet x^2 = 10 \bullet x^5 \bullet x^2 = 10x^{5+2} = 10x^7$

B) $(x^4)^6 = x^{4 \bullet 6} = x^{24}$

C) $\dfrac{8y^8}{2y^2} = \dfrac{8}{2} \bullet \dfrac{y^8}{y^2} = 4 \bullet y^{8-2} = 4y^6$

D) $\left(\frac{x^3}{x^6}\right)\left(\frac{7}{x}\right)^2 = \left(\frac{x^3}{x^6}\right)\left(\frac{7^2}{x^2}\right) = \frac{x^3 \bullet 49}{x^6 \bullet x^2} = \frac{49x^3}{x^{6+2}} = \frac{49x^3}{x^8}$

$$= \frac{49x^3}{x^{5+3}} = \frac{49\cancel{x^3}}{x^5 \bullet \cancel{x^3}} = \frac{49}{x^5}$$

3.2 Radicals

A radical is an expression of the form $\sqrt[n]{a}$ which denotes the nth root of a positive integer a; n is the index of the radical and the number a is the radicand. The index is usually omitted if $n = 2$.

$$\sqrt[2]{a} = \sqrt{a} = a^{1/2}$$

Laws for radicals are the same as laws for exponents, since

$$\sqrt[n]{a} = a^{1/n}, \quad n \neq 0$$

Some of these laws are:

A) $\left(\sqrt[n]{a}\right)^n = a$

B) $\sqrt[n]{ab} = \sqrt[n]{a}\,\sqrt[n]{b}$

C) $\sqrt[n]{a/b} = \frac{\sqrt[n]{a}}{\sqrt[n]{b}}, \quad b \neq 0$

D) $\sqrt[n]{a^m} = \left(\sqrt[n]{a}\right)^m$

E) $\sqrt[m]{\sqrt[n]{a}} = \sqrt[mn]{a}$

A radical is said to be in simplest form if:

A) All perfect nth powers have been removed from the radical.

Example: $\sqrt[3]{8x^5} = \sqrt[3]{(2x)^3 \times x^2} = 2x\left(\sqrt[3]{x^2}\right)$

B) The index of the radical is as small as possible.

C) There aren't any fractions present in the radicand.

Two radicals are said to be similar if they have the same index and the same radicand.

To algebraically add or subtract two or more radicals, reduce each given radical to the simplest form, and add or subtract terms with the same radicals.

Example: $\sqrt{27}+\sqrt{12}=\sqrt{3^2\times 3}+\sqrt{2^2\times 3}=3\sqrt{3}+2\sqrt{3}=5\sqrt{3}$

To multiply two or more radicals with the same radicands, write the radicals in the form a^x, then apply the law

$$a^x a^y = a^{x+y}$$

Example: $\sqrt{2}\times\sqrt[5]{2}\times\sqrt[3]{2^4} = 2^{1/2}\times 2^{1/5}\times 2^{4/3}$

$$=2^{\left(\frac{1}{2}+\frac{1}{5}+\frac{4}{3}\right)}=2^{\frac{61}{30}}=\sqrt[30]{2^{61}}$$

To divide two radicals with the same radicands, write the radicals in the form a^x, then apply the law

$$\frac{a^x}{a^y}=a^{x-y}$$

Example:

$$\left(\sqrt{5}\right)\div\left(\sqrt[3]{5}\right) = \left(5^{\frac{1}{2}}\right)\div\left(5^{\frac{1}{3}}\right)$$

$$= 5^{\left(\frac{1}{2}-\frac{1}{3}\right)}=5^{\frac{1}{6}}$$

$$= \sqrt[6]{5}$$

Problem Solving Examples:

Show that $\sqrt[3]{(-8)^3}=\left(\sqrt[3]{-8}\right)^3$.

$\sqrt[3]{(-8)^3}=\sqrt[3]{-512}=-8.$

Since $(-8)^3 = -512$, $\sqrt[3]{-512} = -8$.

$\sqrt[3]{-8} = -2$ since $(-2)^3 = -8$.

Therefore, $\left(\sqrt[3]{-8}\right)^3 = (-2)^3 = -8$.

$$\sqrt[3]{(-8)^3} = -8 = \left(\sqrt[3]{-8}\right)^3, \text{ hence, } \sqrt[3]{(-8)^3} = \left(\sqrt[3]{-8}\right)^3.$$

Find the indicated roots.

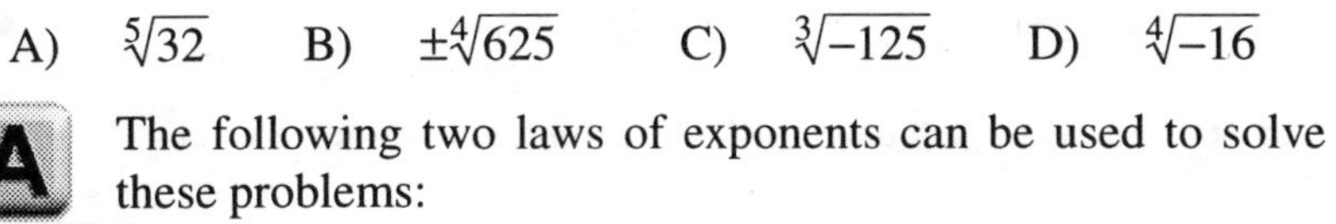

A) $\sqrt[5]{32}$ B) $\pm\sqrt[4]{625}$ C) $\sqrt[3]{-125}$ D) $\sqrt[4]{-16}$

The following two laws of exponents can be used to solve these problems:

1) $\sqrt[n]{a^n} = \left(\sqrt[n]{a}\right)^n = a^1 = a$, and

2) $\sqrt[n]{a^n} = a^{n/n} = a^1 = a$.

A) $\sqrt[5]{32} = \sqrt[5]{2^5} = \left(\sqrt[5]{2}\right)^5 = 2$. This result is true because $(2)^5 = 32$, that is $2 \times 2 \times 2 \times 2 \times 2 = 32$.

B) $\sqrt[4]{625} = \sqrt[4]{5^4} = \left(\sqrt[4]{5}\right)^4 = 5$. This result is true because $(5^4) = 625$, that is, $5 \times 5 \times 5 \times 5 = 625$.

$-\sqrt[4]{625} = -\left(\sqrt[4]{5^4}\right) = -\left[\left(\sqrt[4]{5}\right)^4\right] = -[5] = -5$. This result is true because $(-5)^4 = 625$, that is, $(-5) \times (-5) \times (-5) \times (-5) = 625$.

C) $\sqrt[3]{-125} = \sqrt[3]{(-5)^3} = \left(\sqrt[3]{-5}\right)^3 = -5$. This result is true because $(-5)^3 = -125$, that is, $(-5) \times (-5) \times (-5) = -125$.

D) There is no real solution to $\sqrt[4]{-16}$ because any real number raised to the fourth power is a positive number, that is, $N^4 = (N) \times (N) \times (N) \times (N)$ = a positive number $\neq$ a negative number, -16.

Q Simplify:

A) $\sqrt[3]{-512}$ B) $\sqrt[4]{\frac{81}{16}}$ C) $\sqrt[3]{-16} \div \sqrt[3]{-2}$

A A) By the law of radicals, which states that $\sqrt[n]{ab} = \sqrt[n]{a}\sqrt[n]{b}$, where a and b are any two numbers, $\sqrt[3]{-512} = \sqrt[3]{8(-64)} = \sqrt[3]{8}\sqrt[3]{-64}$. Therefore, $\sqrt[3]{-512} = \sqrt[3]{8}\sqrt[3]{-64} = (2)(-4) = -8$. The last result is true because $(2)^3 = 8$ and $(-4)^3 = -64$.

B) By another law of radicals, which states that $\sqrt[n]{\frac{a}{b}} = \frac{\sqrt[n]{a}}{\sqrt[n]{b}}$, where a and b are any two numbers, $\sqrt[4]{\frac{81}{16}} = \frac{\sqrt[4]{81}}{\sqrt[4]{16}}$. Therefore, $\sqrt[4]{\frac{81}{16}} = \frac{\sqrt[4]{81}}{\sqrt[4]{16}} = \frac{3}{2}$. The last result is true because $(3)^4 = 81$ and $(2)^4 = 16$.

C) By the law of radicals stated in example B), $\sqrt[3]{-16} \div \sqrt[3]{-2} = \frac{\sqrt[3]{-16}}{\sqrt[3]{-2}} = \sqrt[3]{\frac{-16}{-2}} = \sqrt[3]{8} = 2$. The last result is true because $(2)^3 = 8$.

3.3 Scientific Notation

A real number expressed in scientific notation is written as a product of a real number n and an integral power of 10; the value of n ranges from $1 \leq n < 10$.

Example:

	Number	Scientific Notation
1)	1,956.	1.956×10^3

2) .0036 3.6×10^{-3}

3) 5,9600,000. 5.96×10^{7}

Problem Solving Examples:

Use scientific notation to express each number.

A) 4,375

B) 186,000

C) 0.00012

D) 4,005

A number expressed in scientific notation is written as a product of a number between 1 and 10 and a power of 10. The number between 1 and 10 is obtained by moving the decimal point of the number (actual or implied) the required number of digits. The power of 10, for a number greater than 1, is positive and is one less than the number of digits before the decimal point in the original number. The power of 10, for a number less than 1, is negative and is one more than the number of zeros immediately following the decimal point in the original number. Hence,

A) $4{,}375 = 4.375 \times 10^{3}$

B) $186{,}000 = 1.86 \times 10^{5}$

C) $0.00012 = 1.2 \times 10^{-4}$

D) $4{,}005 = 4.005 \times 10^{3}$

Express in scientific notation $\dfrac{6{,}400{,}000}{400}$.

In order to solve this problem, we express the numerator and denominator as the product of a number between 1 and 10 and a power of 10. Thus,

$$6{,}400{,}000 = 6.4 \times 1{,}000{,}000 = 6.4 \times 10^6$$

$$400 = 4 \times 100 = 4 \times 10^2$$

Thus,

$$\frac{6{,}400{,}000}{400} = \frac{6.4 \times 10^6}{4.0 \times 10^2}$$

Since $\frac{ab}{cd} = \frac{a}{c} \times \frac{b}{d}$: $= \frac{6.4}{4.0} \times \frac{10^6}{10^2}$

Since $\frac{a^x}{a^y} = a^{x-y}$: $= 1.6 \times 10^4$

Quiz: Sets and Set Operations – Exponents and Radicals

1. Let $n(A)$ denote the number of elements in set A. If $n(A) = 10$, $n(B) = 12$, and $n(A \cap B) = 3$, how many elements does $A \cup B$ contain?

   (A) 10
   (B) 12
   (C) 15
   (D) 19
   (E) 22

2. If $A \subset C$ and $B \subset C$, which of the following statements are true?

   (A) The set $A \cup B$ is also a subset of C.
   (B) The complement of A is also a subset of C.
   (C) The complement of B is also a subset of C.
   (D) The union of A and B contains C.
   (E) C is the universal set.

3. The General Laws of Exponents are all of the following except:

   (A) $a^p a^q = a^{p+q}$

   (B) $(a^p)^q = a^{pq}$

   (C) $\dfrac{a^p}{a^q} = a^{p-q}, \quad a \neq 0$

   (D) $\left(\dfrac{a}{b}\right)^p = \dfrac{a^p}{b^p}, \quad a = 0$

   (E) $(ab)^p = a^p b^p$

4. If $i = \sqrt{-1}$, then $(a + bi)^2 - (a - bi)^2$ is equivalent to

   (A) $4abi$

   (B) -1

   (C) $a^2 - b^2$

   (D) $2bi$

   (E) $-2b^2$

5. If $x \neq 0$, then $(2^{4x})(8^{2x}) =$

   (A) 16^{3x}

   (B) 2^{10x}

   (C) 4^{4x}

   (D) 2^{16x}

   (E) 16^{2x}

6. If $a^2b^3 = 3^7$ and $a = 9$, then $b =$

   (A) 1

   (B) 2

   (C) $\sqrt{3}$

   (D) 3

   (E) $3\sqrt{3}$

7. $\sqrt{108} + 3\sqrt{12} - 7\sqrt{3} =$

   (A) $3 - 3\sqrt{3}$

   (B) 0

   (C) $4\sqrt{3}$

   (D) $5\sqrt{3}$

   (E) $10\sqrt{3}$

8. $\sqrt{8} + 3\sqrt{18} - 7\sqrt{2} =$

(A) $3 - 3\sqrt{2}$

(B) 0

(C) $6\sqrt{2} - 4\sqrt{3}$

(D) $4\sqrt{2}$

(E) $10\sqrt{2}$

9. $(4d^2 + 7e^3 + 12f) + (3d^2 + 6e^3 + 2f) =$

(A) $23d^2e^3f$

(B) $33d^2e^2f$

(C) $33d^4e^6f^2$

(D) $7d^2 + 13e^3 + 14f$

(E) $23d^2 + 11e^3f$

10. $14m^2n - 6m^2n =$

(A) $20m^2n$

(B) $8m^2n$

(C) $8m$

(D) 8

(E) $8m^4n^2$

ANSWER KEY

1. (D)
2. (A)
3. (D)
4. (A)
5. (B)
6. (D)
7. (D)
8. (D)
9. (D)
10. (B)

4.1 Terms and Expressions

A *variable* is defined as a placeholder which can take on any of several values at a given time; it is usually represented by one of the last letters of the alphabet such as x, y, or z. A *constant* is a symbol which takes on only one value at a given time. If the value of the constant is unknown, it is usually denoted by the letters a, b, or c. π, 5, 8/17, and $-i$ are constants.

A *term* is a constant, a variable, a product of constants and variables, a quotient of constants and variables, or a combination of products and quotients. Terms are separated by addition or subtraction signs. For example: 7.76, $3x$, xyz, $x/5$, $5z/x$, and $(0.99)x^2$ are terms. If a term is a combination of constants and variables, the constant part of the term is referred to as the *coefficient* of the variable. If a variable is written without a coefficient, the coefficient is assumed to be 1.

Example:	$3x^3$	$y^3 = (1)y^3$
	coefficient: 3	coefficient: 1
	variable: x	variable: y

An *expression* is a collection of one or more terms. If the number of terms is greater than 1, the expression is said to be the sum of terms.

Example: $9, 9xy, 6x + x/3, 8yx - 2x$

4.2 The Polynomial

An algebraic expression consisting of only one term is called a *monomial.* An algebraic expression consisting of two terms is called a *binomial.* An algebraic expression consisting of three terms is called a *trinomial.* In general, an algebraic expression consisting of two or more terms is called a *multinomial* or *polynomial.*

A *polynomial* in x, denoted $P(x)$, consists of one or more terms such that the terms are either an integral constant or the product of an integral constant and a positive integral power of x.

Example: $5x^3 + 2x^2 + 3$ is a polynomial in x

$2x^2 + x^{\frac{1}{2}} - 1$ is not a polynomial in x

$9x^3 + 3x^{-2} + 4$ is not a polynomial in x

The degree of a monomial is the sum of the exponents of the variables. The degree of a monomial with no variables is 0.

Example: $5x^2$ has degree 2.

$3x^3y^2x$ has degree 6

9 has degree 0

The degree of a polynomial is equal to the exponent of that term with the highest power of x whose coefficient is not 0.

Example: $5x^4 + 7x + 12$ has degree 4

4.3 Algebraic Operations with Polynomials

Addition of polynomials is achieved by combining like terms, defined as terms which differ only in numerical coefficients.

Example: $(x^2 - 3x + 5) + (4x^2 + 6x - 3)$

Note: Parentheses are used to distinguish polynomials.

By using the commutative and associative laws, we can rewrite $P(x)$ as:

$$P(x) = (x^2 + 4x^2) + (6x - 3x) + (5 - 3)$$

Using the distributive law yields:

$$(1 + 4)x^2 + (6 - 3)x + (5 - 3)$$

$$= 5x^2 + 3x + 2.$$

Subtraction of two polynomials is achieved by first changing the sign of all terms in the expression which is being subtracted and then adding this result to the other expression.

Example: $(5x^2 + 4y^2 + 3z^2) - (4xy + 7y^2 - 3z^2 + 1)$

$$= 5x^2 + 4y^2 + 3z^2 - 4xy - 7y^2 + 3z^2 - 1$$

$$= (5x^2) + (4y^2 - 7y^2) + (3z^2 + 3z^2) - 4xy - 1$$

$$= 5x^2 - 3y^2 + 6z^2 - 4xy - 1$$

Multiplication of two or more monomials is achieved by using the laws of exponents, the rules of signs, and the commutative and associative laws of multiplication.

Example: $(y^2)(5)(6y^2)(yz)(2z^2)$

$$= (1)(5)(6)(1)(2)(y^2)(y^2)(yz)(z^2)$$

$$= (60)[(y^2)(y^2)(y)][(z)(z^2)]$$

$$= 60(y^5)(z^3)$$

$$= 60y^5z^3$$

Multiplication of a polynomial by a monomial is achieved by multiplying each term of the polynomial by the monomials and combining the results (this is the distributive property).

Example: $(4x^2 + 3y) \times (6xz^2)$

$$= 24x^3z^2 + 18xyz^2$$

Multiplication of a polynomial by a polynomial is achieved by multiplying each of the terms of one polynomial by each of the terms of the other polynomial and combining the result.

Example: $(5y + z + 1) \times (y^2 + 2y)$

$$\begin{aligned}
&= [(5y) \times (y^2) + (5y) \times (2y)] \\
&\quad + [(z) \times (y^2) + (z) \times (2y)] \\
&\quad + [(1) \times (y^2) + (1) \times (2y)] \\
&= (5y^3 + 10y^2) + (y^2z + 2yz) + (y^2 + 2y) \\
&= (5y^3) + (10y^2 + y^2) + (y^2z) + (2yz) + (2y) \\
&= 5y^3 + 11y^2 + y^2z + 2yz + 2y.
\end{aligned}$$

Division of a monomial by a monomial is achieved by finding the quotient of the constant coefficients and the quotients of the variable factors, followed by the multiplication of these quotients.

Example: $6xyz^2 \div 2y^2z$

$$\begin{aligned}
&= (6/2)(x/1)(y/y^2)(z^2/z) \\
&= 3xy^{-1}z \\
&= 3xz/y
\end{aligned}$$

Division of a polynomial by a polynomial is achieved by following the given procedure called *long division.*

Step 1: The terms of both the polynomials are arranged in order of ascending or descending powers of one variable.

Step 2: The first term of the dividend is divided by the first term of the divisor which gives the first term of the quotient.

Step 3: The divisor is multiplied by the first term of the quotient and the result is subtracted from the dividend.

Step 4: Using the remainder obtained from step 3 as the new dividend, steps 2 and 3 are repeated until the remainder is zero or the degree of the remainder is less than the degree of the divisor.

Step 5: The result is written as follows:

$$\frac{\text{dividend}}{\text{divisor}} = \text{quotient} + \frac{\text{remainder}}{\text{divisor}}$$

divisor $\neq 0$

Example: $(2x^2 + x + 6) \div (x + 1)$

$$\begin{array}{r}
2x - 1 \\
x+1\overline{)2x^2 + x + 6} \\
-(2x^2 + 2x) \\
\hline
-x + 6 \\
-(-x - 1) \\
\hline
7
\end{array}$$

The result is

$$(2x^2 + x + 6) \div (x + 1) = 2x - 1 + \frac{7}{x+1}$$

Problem Solving Examples:

Add $(3xy^2 + 2xy + 5x^2y) + (2xy^2 - 4xy + 2x^2y)$.

Use the vertical form, align all like terms, and apply the distributive property.

$$\begin{array}{c}
3xy^2 + 2xy + 5x^2y \\
2xy^2 - 4xy + 2x^2y \\
\hline
(3 + 2)xy^2 + (2 - 4)xy + (5 + 2)x^2y
\end{array}$$

Thus, the sum is $5xy^2 - 2xy + 7x^2y$.

Subtract $4y^2 - 5y + 2$ from $7y^2 - 6$.

The problem is the following:

$$(7y^2 - 6) - (4y^2 - 5y + 2).$$

Whenever a minus sign (–) appears before an expression in parentheses, change the sign of every term in the parentheses.

In this problem, since a minus sign appears before the expression $(4y^2 - 5y + 2)$, the sign of every term in this expression is changed.

$$(4y^2 - 5y + 2) \text{ becomes } -4y^2 + 5y - 2.$$

After the signs have been changed, the new expression $-4y^2 + 5y - 2$ can be added to the expression $7y^2 - 6$ (changing the signs of the expression changes the original problem to an addition problem). Therefore,

$$(7y^2 - 6) - (4y^2 - 5y + 2) = 7y^2 - 6 - 4y^2 + 5y - 2$$

Place the terms with similar powers together. Therefore, the above becomes:

$$\begin{aligned}(7y^2 - 6) - (4y^2 - 5y + 2) &= 7y^2 - 6 - 4y^2 + 5y - 2 \\ &= 7y^2 - 4y^2 + 5y - 6 - 2\end{aligned}$$

Since $7y^2 - 4y^2 = 3y^2$ and $-6 - 2 = -8$, the above becomes:

$$(7y^2 - 6) - (4y^2 - 5y + 2) = 3y^2 + 5y - 8,$$

which is the final answer.

Find the product $(2x - 5y)(x + 2y)$.

We use the distributive property and simplify.

$$\begin{aligned}(2x - 5y)(x + 2y) &= (2x - 5y)x + (2x - 5y)2y && \text{Distributive property} \\ &= [2x + x + (-5y)\,x] + && \text{Distributive property} \\ &\quad [2x + 2y + (-5y) \times 2y]\end{aligned}$$

$$= 2x^2 - 5xy + 4xy - 10y^2 \qquad \text{Simplifying}$$

$$= 2x^2 - xy - 10y^2 \qquad \text{Combining like terms}$$

So,

$$(2x - 5y)(x + 2y) = 2x^2 - xy - 10y^2$$

Q Divide $3x^5 - 8x^4 - 5x^3 + 26x^2 - 33x + 26$ by $x^3 - 2x^2 - 4x + 8$.

A To divide a polynomial by another polynomial, we set up the divisor and the dividend as shown below. Then we divide the first term of the divisor into the first term of the dividend. We multiply the quotient from this division by each term of the divisor, and subtract the products of each term from the dividend. We then obtain a new dividend. Use this dividend and again divide by the first term of the divisor. Repeat all steps again until we obtain a remainder which is of a degree lower than that of the divisor or = zero. Following this procedure we obtain:

$$\begin{array}{r}
3x^2 - 2x + 3 \\
x^3 - 2x^2 - 4x + 8 \overline{\smash{)}\, 3x^5 - 8x^4 - 5x^3 + 26x^2 - 33x + 26} \\
-(3x^5 - 6x^4 - 12x^3 + 24x^2) \\
\hline
-2x^4 + 7x^3 + 2x^2 - 33x \\
-(-2x^4 + 4x^3 + 8x^2 - 16x) \\
\hline
3x^3 - 6x^2 - 17x + 26 \\
-(3x^3 - 6x^2 - 12x + 24) \\
\hline
-5x + 2
\end{array}$$

Thus, the quotient is $3x^2 - 2x + 3$ and the remainder is $-5x + 2$.

4.4 Polynomial Factorization

To factor a polynomial completely is to find the prime factors of the polynomial with respect to a specified set of numbers, i.e., to express it as a product of polynomials whose coefficients are members of that set.

The following concepts are important while factoring polynomials:

The factors of an algebraic expression consist of two or more algebraic expressions which when multiplied together produce the given algebraic expression.

A prime factor is a polynomial with no factors other than itself and 1. The least common multiple for a set of numbers is the smallest quantity divisible by every number of the set. For algebraic expressions, the least common multiple is the polynomial of lowest degree and smallest numerical coefficients for which each of the given expressions will be a factor.

The greatest common factor for a set of numbers is the largest factor that is common to all members of the set.

For algebraic expressions, the greatest common factor is the polynomial of highest degree and largest numerical coefficients which is a factor of all the given expressions.

Some important formulas, useful for the factoring of polynomials, are listed below.

$$a(c + d) = ac + ad$$

$$(a + b)(a - b) = a^2 - b^2$$

$$(a + b)(a + b) = (a + b)^2 = a^2 + 2ab + b^2$$

$$(a - b)(a - b) = (a - b)^2 = a^2 - 2ab + b^2$$

$$(x + a)(x + b) = x^2 + (a + b)x + ab$$

$$(ax + b)\,(cx + d) = acx^2 + (ad + bc)x + bd$$

$$(a + b)(c + d) = ac + bc + ad + bd$$

$$(a + b)(a + b)(a + b) = (a + b)^3 = a^3 + 3a^2b + 3ab^2 + b^3$$

$(a - b)(a - b)(a - b) = (a - b)^3 = a^3 - 3a^2b + 3ab^2 - b^3$

$(a + b)(a^2 - ab + b^2) = a^3 + b^3$

$(a + b + c)^2 = a^2 + b^2 + c^2 + 2ab + 2ac + 2bc$

$(a - b)(a^2 + ab + b^2) = a^3 - b^3$

$(a - b)(a^3 + a^2b + ab^2 + b^3) = a^4 - b^4$

$(a - b)(a^4 + a^3b + a^2b^2 + ab^3 + b^4) = a^5 - b^5$

$(a - b)(a^5 + a^4b + a^3b^2 + a^2b^3 + ab^4 + b^5) = a^6 - b^6$

$(a - b)(a^{n-1} + a^{n-2}b + a^{n-3}b^2 + \ldots + ab^{n-2} + b^{n-1}) = a^n - b^n$

where n is any positive integer (1, 2, 3, 4, ...).

$(a + b)(a^{n-1} - a^{n-2}b + a^{n-3}b^2 - \ldots - ab^{n-2} + b^{n-1}) = a^n + b^n$

where n is any positive odd integer (1, 3, 5, 7, ...).

The procedure for factoring a polynomial completely is as follows:

Step 1: First, find the greatest common factor if there is any. Then, by grouping terms, examine each group for greatest common factors.

Step 2: Continue factoring the factors obtained in step 1 until all factors other than monomial factors are prime.

Example: Factoring $4 - 16x^2$,

$$4 - 16x^2 = 4(1 - 4x^2) = 4(1 + 2x)(1 - 2x)$$

Problem Solving Examples:

Factor the following polynomials:

A) $15ac + 6bc - 10ad - 4bd$

B) $3a^2c + 3a^2d^2 + 2b^2c + 2b^2d^2$

A) Group terms that have common factors. For example:

$(15ac + 6bc) - (10ad + 4bd)$

Then factor.

$$(15ac + 6bc) - (10ad - 4bd) = 3c(5a + 2b) - 2d(5a + 2b)$$

Factoring out $(5a + 2b)$,

$$3c(5a + 2b) - 2d(5a + 2b) \quad = (5a + 2b)(3c - 2d)$$

B) Apply the same method as in A),

$$\begin{aligned} 3a^2c + 3a^2d^2 + 2b^2c + 2b^2d^2 &= (3a^2c + 3a^2d^2) + (2b^2c + 2b^2d^2) \\ &= 3a^2(c + d^2) + 2b^2(c + d^2) \\ &= (c + d^2)(3a^2 + 2b^2) \end{aligned}$$

Q Factor $xy - 3y + y^2 - 3x$ completely.

A Note that the first and last terms have a common factor of x. Also note that the second and third terms have a common factor of y. Hence, group the x and y terms together and factor out the x and y from their respective two terms. Therefore,

$$xy - 3y + y^2 - 3x = (xy - 3x) + (-3y + y^2)$$

Since $$(-3y + y^2) = (y^2 - 3y),$$

$$\begin{aligned} xy - 3y + y^2 - 3x &= (xy - 3x) + (y^2 - 3y) \\ &= x(y - 3) + y(y - 3) \end{aligned}$$

Now factor out the common factor $(y - 3)$ from both terms:

$$xy - 3y + y^2 - 3x = (x + y)(y - 3).$$

4.5 Operations with Fractions and Rational Expressions

A *rational* expression is an algebraic expression that can be written as the quotient of two polynomials, $\frac{A}{B}$, $B \neq 0$.

Example: $\frac{9}{4}, \frac{3x^2+5x}{y+3}, \frac{9y}{10z}$

To reduce a given fraction or a rational expression to its simplest form is to reduce the fraction or expression into an equivalent form such that its numerator and denominator have no common factor other than 1.

The operations performed on fractions are in a similar manner applicable to rational expressions.

A fraction that contains one or more fractions in either its numerator or denominator—or in both its numerator and denominator—is called a complex fraction.

Example: $\frac{\frac{1}{x}}{\frac{5}{y}}, \frac{\frac{1}{2}}{3}, \frac{1+\frac{y}{x}}{1-\frac{4}{x^2+1}}$

The procedure for simplifying complex fractions is as follows:

First, the terms in the numerator and denominator are separately combined. Then the combined term of the numerator is divided by the combined term of the denominator to obtain a simplified fraction.

Example: $\frac{1-\frac{5}{x}+\frac{6}{x^2}}{1-\frac{6}{x}+\frac{8}{x^2}}$

Combining the numerator, we get

$$1-\frac{5}{x}+\frac{6}{x^2} = \frac{x^2-5x+6}{x^2}$$

$$= \frac{(x-3)(x-2)}{x^2}$$

Combining the denominator we get

$$1-\frac{6}{x}+\frac{8}{x^2} = \frac{x^2-6x+8}{x^2}$$

$$= \frac{(x-4)(x-2)}{x^2}$$

Dividing the resultant numerator by the resultant denominator, we get:

$$\frac{\frac{(x-3)(x-2)}{x^2}}{\frac{(x-4)(x-2)}{x^2}} = \frac{(x-3)(x-2)}{x^2} \times \frac{x^2}{(x-4)(x-2)}$$

$$= \frac{x-3}{x-4}$$

Problem Solving Example:

Simplify:

$$\frac{\frac{1}{a-b}+\frac{1}{a+b}}{1+\frac{b^2}{a^2-b^2}}$$

A This is a complex fraction, a fraction whose numerator and denominator both contain fractions. To simplify it, multiply the numerator and denominator by the least common denominator, LCD. To find the LCD of several fractions, first factor each denominator into its prime factors.

$$a - b = (a - b)$$

$$a + b = (a + b)$$

$$a^2 - b^2 = (a - b)(a + b)$$

The LCD of the fractions is the product of the highest power of the different prime factors, with each prime factor being used only once. Hence, $(a - b)(a + b)$ is our LCD. Multiplying, we obtain:

$$\frac{(a-b)(a+b)\left[\frac{1}{a-b}+\frac{1}{a+b}\right]}{(a-b)(a+b)\left[1+\frac{b^2}{a^2-b^2}\right]}$$

Distributing in the numerator and denominator, and recalling that $a^2 - b^2 = (a - b)(a + b)$, we have:

$$\frac{\frac{\cancel{(a-b)}(a+b)}{\cancel{(a-b)}}+\frac{(a-b)\cancel{(a+b)}}{\cancel{(a+b)}}}{(a-b)(a+b)+\frac{b^2\cancel{(a-b)}\cancel{(a+b)}}{\cancel{(a-b)}\cancel{(a+b)}}}=\frac{(a+b)+(a-b)}{(a-b)(a+b)+b^2}$$

$$=\frac{a+b+a-b}{a^2-b^2+b^2}=\frac{2a}{a^2}=\frac{2}{a}.$$

Quiz: Polynomials and Rational Expressions

1. $(2a + b)(3a^2 + ab + b^2) =$

   (A) $6a^3 + 5a^2b + 3ab^2 + b^3$

   (B) $5a^3 + 3ab + b^3$

   (C) $6a^3 + 2a^2b + 2ab^2$

   (D) $3a^2 + 2a + ab + b + b^2$

   (E) $6a^3 + 3a^2b + 5ab^2 + b^3$

2. $(x^2 + x - 6) \div (x - 2) =$

   (A) $x - 3$

   (B) $x + 2$

   (C) $x + 3$

   (D) $x - 2$

   (E) $2x + 2$

3. $x^2 + xy - 2y^2 =$

(A) $(x - 2y)(x + y)$
(B) $(x - 2y)(x - y)$
(C) $(x + 2y)(x + y)$
(D) $(x + 2y)(x - y)$
(E) Not possible.

4. Simplify $\dfrac{4x^3 + 6x^2}{2x}$.

(A) $2x^2 + 3$
(B) $8x^2$
(C) $2x + 3$
(D) $2x^3 + 3x^2$
(E) $2x^2 + 3x$

5. Factor $x^2 + 7x + 12$.

(A) $(x + 2)(x + 5)$
(B) $(x + 2)(x + 6)$
(C) $(x - 4)(x + 3)$
(D) $(x + 4)(x + 3)$
(E) Cannot be factored.

6. Simplify $\dfrac{x^2 - 3x}{x^2 - 9} + \dfrac{3}{x + 3}$.

(A) $x + 3$
(B) $x - 3$
(C) $\dfrac{1}{3}$
(D) $x^2 - 3x + 3$
(E) 1

7. Add the expressions

$4a^2 - 3 + 5a$,
$6a - 2a^2 + 2$, and
$2a^2 - 3a + 8$.

(A) $12a + 7$

(B) $12a^2 - 8a + 15$

(C) $4a^2 + 8a + 7$

(D) $8a^2 - 12a + 13$

(E) 19

8. Simplify $\dfrac{x^2 - y^2}{x + y}$.

(A) xy

(B) 0

(C) $\dfrac{2(x-y)}{x+y}$

(D) $x - y$

(E) $x + y$

9. Subtract

$3x^4y^3 + 5x^2y - 4xy + 5x - 3$ from $5x^4y^3 - 3x^2y + 7$

(A) $-2x^4y^3 + 8x^2y - 4xy + 5x - 10$

(B) $2x^4y^3 + 2x^2y - 4xy + 5x + 4$

(C) $-16x^6y^4 - 20x^2y + 10$

(D) $2x^4y^3 - 8x^2y + 4xy - 5x + 10$

(E) $-2x^4y^3 - 2x^2y - 4xy + 5x + 4$

10. Find the product $(2x^2 - 3xy + y^2)(2x - y)$.

(A) $4x^3 - 8x^2y + 5xy^2 - y^3$

(B) $4x^3 + 4x^2y + 51xy^2 - y^3$

(C) $-2x^4y + x^3y^2 + 2xy^2 - y^3$

(D) $-4x^3y + 6x^2y^2 - 4xy^3$

(E) $4x^3 - 3x^2y^2 - y^3$

ANSWER KEY

1. (A)
2. (C)
3. (D)
4. (E)
5. (D)
6. (E)
7. (C)
8. (D)
9. (D)
10. (A)

CHAPTER 5

Equations

5.1 Equations

An equation is defined as a statement of equality of two separate expressions known as members.

A conditional equation is an equation that is true for only certain values of the unknowns (variables) involved.

Example: $y + 6 = 11$ is true for $y = 5$

An equation that is true for all permissible values of the unknown in question is called an identity. For example, $2x = \frac{4}{2}x$ is an identity of $x \in R$, i.e., it is true for all reals.

The values of the variables that satisfy a conditional equation are called solutions of the conditional equation; the set of all such values is known as the solution set.

The solution to an equation $f(x) = 0$ is called the root of the equation.

Equations with the same solutions are said to be equivalent equations.

A statement of equality between two expressions containing rational coefficients and whose exponents are integers is called a rational integral equation. The degree of the equation is given by the term with highest power, as shown below:

$$a_n x^n + a_{n-1} x^{n-1} + a_{n-2} x^{n-2} + \ldots + a_1 x + a_0 = 0$$

where $a_n \neq 0$, the a_1, $i = 1 \ldots n$ are rational constant coefficients and n is a positive integer.

5.2 Basic Laws of Equality

A) Replacing an expression of an equation by an equivalent expression results in an equation equivalent to the original one.

Example: Given the equation below

$$3x + y + x + 2y = 15$$

We know that for the left side of this equation we can apply the commutative and distributive laws to get:

$$3x + y + x + 2y = 4x + 3y.$$

Since these are equivalent, we can replace the expression in the original equation with the simpler form to get:

$$4x + 3y = 15.$$

B) The addition or subtraction of the same expression on both sides of an equation results in an equation equivalent to the original one.

Example: Given the equation

$$y + 6 = 10,$$

we can add (–6) to both sides

$$y + 6 + (-6) = 10 + (-6)$$

to get

$$y + 0 = 10 - 6 \Rightarrow y = 4.$$

So $y + 6 = 10$ is equivalent to $y = 4$.

C) The multiplication or division on both sides of an equation by the same expression results in an equation equivalent to the original.

Example: $3x = 6 \Rightarrow \frac{3x}{3} = \frac{6}{3} \Rightarrow x = 2$

$3x = 6$ is equivalent to $x = 2$.

D) If both members of an equation are raised to the same power, then the resultant equation is equivalent to the original equation.

Example: If $a = x^2y$, then $(a)^2 = (x^2y)^2 \Rightarrow a^2 = x^3y^2$.

This applies for negative and fractional powers as well.

Example: $x^2 = 3y^4$

If we raise both members to the –2 power we get

$$(x^2)^{-2} = (3y^4)^{-2}$$

$$\frac{1}{(x^2)^2} = \frac{1}{(3y^4)^2}$$

$$\frac{1}{x^4} = \frac{1}{9y^8}$$

If we raise both members to the $\frac{1}{2}$ power, which is the same as taking the square root, we get:

$$(x^2)^{\frac{1}{2}} = (3y^4)^{\frac{1}{2}}$$

$$x = \pm\sqrt{3}y^2$$

E) The reciprocal of both members of an equation are equivalent to the original equation.

Note: The reciprocal of zero is undefined.

$$\frac{2x+y}{z} = \frac{5}{2} \qquad \frac{z}{2x+y} = \frac{2}{5}$$

Problem Solving Examples:

Solve $3x - 5 = 4$ for x.

Since $3x - 5 = 4$ is to be solved for x, isolate x on one side of the equation. First, add 5 to both sides of the equation.

$$3x - 5 + 5 = 4 + 5 \qquad (1)$$

Since $-5 + 5 = 0$ and $4 + 5 = 9$, equation (1) reduces to:

$$3x = 9. \qquad (2)$$

Since it is desired to get the term x isolated on one side of the equation, divide both sides of Equation (2) by 3.

$$\frac{3x}{3} = \frac{9}{3} \qquad (3)$$

Since $\frac{3x}{3}$ reduces to $1x$, and since $\frac{9}{3}$ reduces to 3, Equation (3) becomes: $1x = 3$. Since $1x = x$, $x = 3$. Therefore, the equation has been solved for x.

Check: By substituting $x = 3$ into the original equation, we have

$$3(3) - 5 \stackrel{?}{=} 4$$
$$9 - 5 \stackrel{?}{=} 4$$
$$4 = 4$$

Note that, upon substitution of the solution into the original equation, the equation is reduced to the identity $4 = 4$.

Solve the equation $6x - 3 = 7 + 5x$.

To solve for x in the equation $6x - 3 = 7 + 5x$, we wish to obtain an equivalent equation in which each term in one member involves x, and each term in the other member is a constant. If we add $(-5x)$ to both members, then only one side of the equation will have an x term:

$$6x - 3 + (-5x) = 7 + 5x + (-5x)$$
$$6x + (-5x) - 3 = 7 + 0$$
$$x - 3 = 7$$

Now, adding 3 to both sides of the equation, we obtain,

$$x - 3 + 3 = 7 + 3$$
$$x + 0 = 10$$
$$x = 10$$

Thus, our solution is $x = 10$. Now we check this value.

Check: Substitute 10 for x in the original equation:

$$6x - 3 = 7 + 5x$$
$$6(10) - 3 = 7 + 5(10)$$
$$60 - 3 = 7 + 50$$
$$57 = 57$$

Solve, justifying each step: $3x - 8 = 7x + 8$.

A

	$3x - 8 = 7x + 8$
Adding 8 to both members	$3x - 8 + 8 = 7x + 8 + 8$
Additive inverse property	$3x + 0 = 7x + 16$
Additive identity property	$3x = 7x + 16$
Adding $(-7x)$ to both members	$3x - 7x = 7x + 16 - 7x$
Commuting	$-4x = 7x - 7x + 16$
Additive inverse property	$-4x = 0 + 16$
Additive identity property	$-4x = 16$
Dividing both sides by -4	$\frac{-4x}{-4} = \frac{16}{-4}$
	$x = -4$

Check: Replacing x by -4 in the original equation:

$$3x - 8 = 7x + 8$$
$$3(-4) - 8 = 7(-4) + 8$$
$$-12 - 8 = -28 + 8$$
$$-20 = -20$$

5.3 Equations with Absolute Values

When evaluating equations containing absolute values, proceed as follows:

Example: $| 5 - 3x | = 7$ is valid if either

$5 - 3x = 7$	or	$5 - 3x = -7$
$-3x = 2$		$-3x = -12$
$x = \frac{-2}{3}$		$x = 4$

The solution set is therefore $x = (-2/3, 4)$, since those values will solve the equation as either 7 or –7.

Problem Solving Examples:

Solve for x when $| x - 7 | = 3$.

This equation, according to the definition of absolute value, expresses the condition that $x - 7$ must be 3 or –3, since in either case the absolute value is 3. If $x - 7 = 3$, we have $x = 10$; and if $x - 7 = -3$, we have $x = 4$. We see that there are two values of x which solve the equation.

Solve for x when $| 5x + 4 | = -3$.

In examining the given equation, it is seen that the absolute value of a number is set equal to a negative value. By definition of an absolute value, however, the number cannot be negative. Therefore, the given equation has no solution.

CHAPTER 6

Linear Equations and Systems of Linear Equations

6.1 Linear Equations

A linear equation in one variable is one that can be put into a form such as $ax + b = 0$, where a and b are constants, $a \neq 0$.

To solve a linear equation means to transform it to the form $x = \frac{-b}{a}$.

A) If the equation has unknowns on both sides of the equality, it is convenient to put similar terms on the same side.

Example:

$$4x + 3 = 2x + 9$$
$$4x + 3 - 2x = 2x + 9 - 2x$$
$$(4x - 2x) + 3 = (2x - 2x) + 9$$
$$2x + 3 = 0 + 9$$
$$2x + 3 - 3 = 0 + 9 - 3$$
$$2x = 6$$
$$\frac{2x}{2} = \frac{6}{2}$$
$$x = 3$$

B) If the equation appears in fractional form, it is necessary to transform it, using cross-multiplication, and then repeating the same procedure as in A). We obtain:

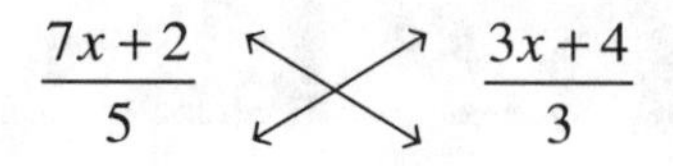

By using cross-multiplication we would obtain:

$$3(7x + 2) = 5(3x + 4).$$

This is equivalent to:

$$21x + 6 = 15x + 20,$$

which can be solved as in A):

$$21x + 6 = 15x + 20$$
$$21x - 15x + 6 = 15x - 15x + 20$$
$$6x + 6 - 6 = 20 - 6$$
$$6x = 14$$
$$x = \frac{14}{6}$$
$$x = \frac{7}{3}$$

C) If there are radicals in the equation, it is necessary to square both sides and then apply A):

$$\sqrt{3x+1} = 5$$
$$\left(\sqrt{3x+1}\right)^2 = 5^2$$
$$3x + 1 = 25$$
$$3x + 1 - 1 = 25 - 1$$
$$3x = 24$$
$$x = \frac{24}{3}$$
$$x = 8$$

Problem Solving Examples:

Solve the equation $\frac{3}{4}x+\frac{7}{8}+1=0$.

There are several ways to proceed. First, we observe that

$$\frac{3}{4}x+\frac{7}{8}+1=0$$

is equivalent to

$$\frac{3}{4}x+\frac{7}{8}+\frac{8}{8}=0$$

where we have converted 1 into $\frac{8}{8}$. Now, combining fractions we obtain:

$$\frac{3}{4}x+\frac{15}{8}=0.$$

Subtract $\frac{15}{8}$ from both sides:

$$\frac{3}{4}x=\frac{-15}{8}.$$

Multiply both sides by $\frac{4}{3}$:

$$\left(\frac{4}{3}\right)\frac{3}{4}x=\left(\frac{4}{3}\right)\left(\frac{-15}{8}\right).$$

Cancel like terms in the numerator and denominator:

$$x=\frac{-5}{2}.$$

A second method is to multiply both sides of the equation by the least common denominator, 8:

$$8\left(\frac{3}{4}x+\frac{7}{8}+1\right)=8\ (0).$$

Distribute: $8\left(\frac{3}{4}\right)x+8\left(\frac{7}{8}\right)+8\times1=0$

$$(2\times3)x+7+8=0$$

$$6x+15=0.$$

Subtract 15 from both sides: $6x=-15.$

Divide both sides by 6: $x=\frac{-15}{6}.$

Cancel 3 from numerator and denominator: $x=\frac{-5}{2}.$

Solve the equation $\frac{3}{2}x-\frac{2}{3}=2x+1.$

Subtract $\frac{3}{2}x$ from both sides of the given equation:

$$\cancel{\frac{3}{2}}x-\frac{2}{3}-\cancel{\frac{3}{2}}x=2x+1-\frac{3}{2}x$$

$$-\frac{2}{3}=\frac{4}{2}x-\frac{3}{2}x+1$$

$$-\frac{2}{3}=\frac{1}{2}x+1$$

$$-\frac{2}{3}=\frac{x}{2}+1.$$

Subtract 1 from both sides of this equation:

$$-\frac{2}{3}-1=\frac{x}{2}+\cancel{1}-\cancel{1}$$

$$-\frac{2}{3}-\frac{3}{3}=\frac{x}{2}$$

$$-\frac{5}{3}=\frac{x}{2}.$$

Multiply both sides of this equation by 2:

$$2\left(-\frac{5}{3}\right) = 2\left(\frac{x}{2}\right)$$

$$-\frac{10}{3} = x.$$

Thus, the solution set of our given equation is the set $\left\{-\frac{10}{3}\right\}$.

6.2 Linear Equations in Two Variables

Equations of the form $ax + by = c$, where a, b, and c are constants and $a, b \neq 0$ are called linear equations in two variables. This equation is also known as the general form of a linear equation.

The solution set for a linear equation in two variables is the set of all x and y values for which the equation is true. An element in the solution set is called an ordered pair (x, y) where x and y are two of the values that together satisfy the equation. The x value is always first and is called the x-coordinate. The y value is always second and is called the y-coordinate.

6.3 Graphing the Solution Set

The solution set of the equation $ax + by = c$ can be represented by graphing the ordered pairs that satisfy the equation on a rectangular coordinate system. This is a system where two real number lines are drawn at right angles to each other. The x-axis is the horizontal line and the y-axis is the vertical line. The point where the two lines intersect is called the origin and is associated to the ordered pair $(0, 0)$.

To plot a certain ordered pair (x, y), move x units along the x-axis in the direction indicated by the sign of x, then move y units along the y-axis in the direction indicated by the sign of y. Note that movement to the right or up is positive, while movement to the left or down is negative.

Example: Graph the following points: (1, 2), (–3, 2), (–2, –1), (1, –1). See Figure 6.1.

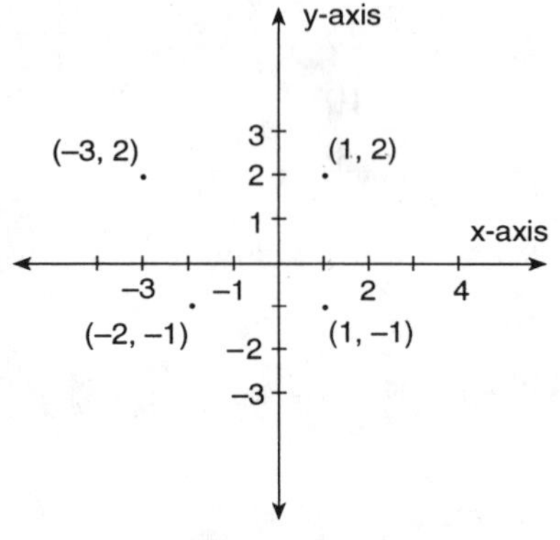

Figure 6.1

To graph a linear equation in two variables, it is necessary to graph its solution set; that is, draw a line through the points whose coordinates satisfy the equation. The resultant graph of a linear equation in two variables is a straight line.

There are several ways of graphing this line (see Figure 6.2), two of them are shown below:

A) Plot two or more ordered pairs that satisfy the equation and then draw the straight line through these points.

B) Plot the points $A\left(\frac{c}{a}, 0\right)$ and $B\left(0, \frac{c}{b}\right)$ that correspond to the points where the line intersects the x-axis and y-axis, respectively, as shown:

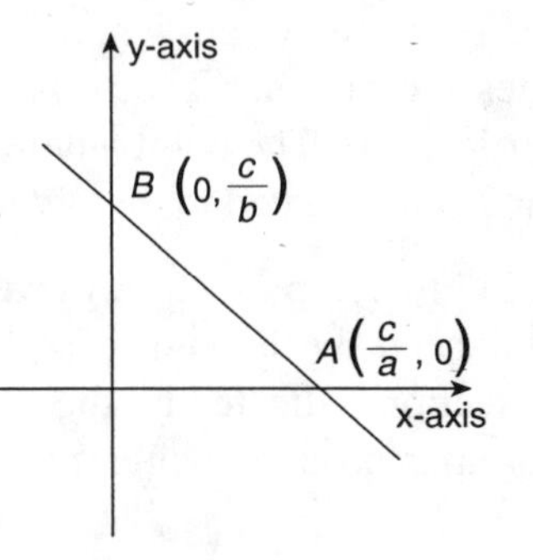

Figure 6.2

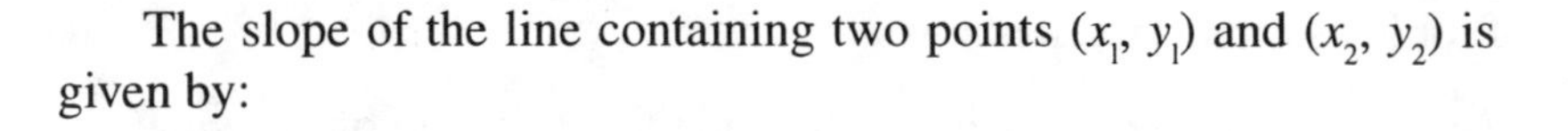

The slope of the line containing two points (x_1, y_1) and (x_2, y_2) is given by:

$$\text{Slope} = m = \frac{y_2 - y_1}{x_2 - x_1}$$

Horizontal lines have a slope of zero, and the slope of vertical lines is undefined. Parallel lines have equal slopes and perpendicular lines have slopes that are negative reciprocals of each other.

The equation of a line with slope m passing through a point $Q(x_0, y_0)$ is of the form:

$$y - y_0 = m(x - x_0)$$

This is called the *point-slope form* of a linear equation.

The equation of a line passing through $Q(x_1, y_1)$ and $P(x_2, y_2)$ is given by:

$$\frac{y - y_1}{x - x_1} = \frac{y_2 - y_1}{x_2 - x_1}$$

This is the *two-point form* of a linear equation.

The equation of a line intersecting the x-axis at $(x_0, 0)$ and the y-axis at $(0, y_0)$ is given by:

$$\frac{x}{x_0} + \frac{y}{y_0} = 1$$

This is the *intercept form* of a linear equation.

The equation of a line with slope m intersecting the y-axis at $(0, b)$ is given by:

$$y = mx + b$$

This is the *slope-intercept* form of a linear equation.

Problems on linear equations:

A) Find the slope, the y-intercept, and the x-intercept of the equation $2x - 3y - 18 = 0$.

Solution: The equation $2x - 3y - 18 = 0$ can be written in the form of the general linear equation, $ax + by = c$.

$$2x - 3y - 18 = 0$$

$$2x - 3y = 18$$

To find the slope and y-intercept, we derive them from the formula of the general linear equation $ax + by = c$. Dividing by b and solving for y, we obtain:

$$\frac{a}{b}x + y = \frac{c}{b}$$

$$y = \frac{c}{b} - \frac{a}{b}x$$

where $\frac{-a}{b}$ = slope and $\frac{c}{b}$ = y-intercept.

To find the x-intercept, solve for x and let $y = 0$:

$$x = \frac{c}{a} - \frac{b}{a}y$$

$$x = \frac{c}{a}$$

In this form we have $a = 2$, $b = -3$, and $c = 18$. Thus,

$$\text{slope} = -\frac{a}{b} = -\frac{2}{-3} = \frac{2}{3}$$

$$y\text{-intercept} = \frac{c}{b} = \frac{18}{-3} = -6$$

$$x\text{-intercept} = \frac{c}{a} = \frac{18}{2} = 9$$

B) Find the equation for the line passing through (3, 5) and (–1, 2).

Solution A): We use the two-point form with $(x_1, y_1)=(3, 5)$ and $(x_2, y_2) = (-1, 2)$. Then

$$\frac{y - y_1}{x - x_1} = \frac{y_2 - y_1}{x_2 - x_1}$$

$$\frac{y_2 - y_1}{x_2 - x_1} = \frac{2-5}{-1-3} \quad \text{thus} \quad \frac{y-5}{x-3} = \frac{-3}{-4}$$

Cross multiply, $-4(y - 5) = -3(x - 3)$.

Distribute, $-4y + 20 = -3x + 9$

Place in general form, $3x - 4y = -11$.

Solution B): Does the same equation result if we let $(x_1, y_1) = (-1, 2)$ and $(x_2, y_2) = (3, 5)$?

$$\frac{y_2 - y_1}{x_2 - x_1} = \frac{5-2}{3-(-1)} \quad \text{thus} \quad \frac{y-2}{x+1} = \frac{3}{4}$$

Cross multiply, $4(y - 2) = 3(x + 1)$

Distribute $3x - 4y = -11$

Place in general form, $3x - 4y = -11$.

Hence, either replacement results in the same equation. Keep in mind that the coefficient of the x-term should always be positive.

C) (a) Find the equation of the line passing through (2,5) with slope 3.

(b) Suppose a line passes through the y-axis at (0,b). How can we write the equation if the point-slope form is used?

Solution C): (a) In the point-slope form, let $x_1 = 2$, $y_1 = 5$, and $m = 3$. The point-slope form of a line is:

$$y - y_1 = m(x - x_1)$$

$$y - 5 = 3(x - 2)$$

$$y - 5 = 3x - 6 \quad \text{Distributive property}$$

$$y = 3x - 1 \quad \text{Transposition}$$

(b) $y - b = m(x - 0)$

$$y = mx + b.$$

Notice that this is the slope-intercept form for the equation of a line.

Problem Solving Examples:

Construct the graph of the function defined by $y = 3x - 9$.

This linear equation is in the slope-intercept form, $y = mx + b$.

A line can be determined by two points. Let us choose the intercepts. The x-intercept lies on the x-axis and the y-intercept is on the y-axis.

We can find the y-intercept by assigning 0 to x in the given equation and then find the x-intercept by assigning 0 to y. It is helpful to have a third point. We find a third point by assigning 4 to x and solving for y. Thus, we get the following table of corresponding numbers:

x	$y = 3x - 9$	y
0	$y = 3(0) - 9$	-9
3	$0 = 3x - 9, x = 9/3 = 3$	0
4	$y = 3(4) - 9$	3

The three points are (0, –9), (3, 0), and (4, 3). Draw a line through them (see Figure 6.3).

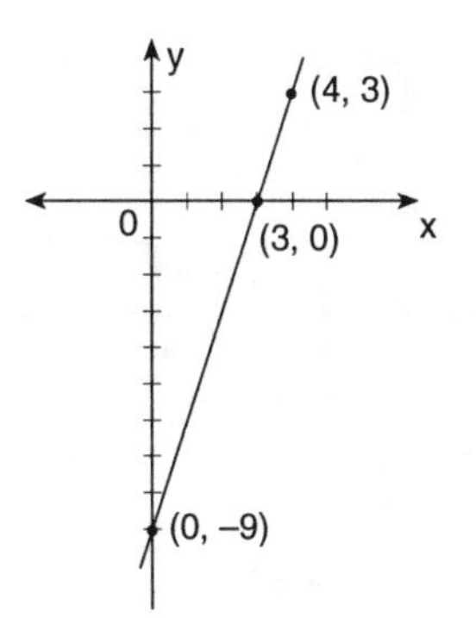

Figure 6.3

Graph the function defined by $3x - 4y = 12$.

Solve for y:

$$\begin{aligned} 3x - 4y &= 12 \\ -4y &= 12 - 3x \\ y &= -3 + \frac{3}{4}x \\ y &= \frac{3}{4}x - 3. \end{aligned}$$

The graph of this function is a straight line since it is of the form $y = mx + b$. The y-intercept crosses (intersects) the y-axis at the point (0, –3) since for $x = 0$, $y = b = -3$. The x-intercept crosses (intersects) the x-axis at the point (4, 0) since for $y = 0$, $x = (y+3)\times\frac{4}{3} = (0+3)\times\frac{4}{3} = 4$. These two points, (0, –3) and (4, 0) are sufficient to determine the graph (see Figure 6.4). A third point, (8, 3), satisfying the equation of the function is plotted as a partial check of the intercepts. Note that the slope of the line is $m = \frac{3}{4}$. This means that y increases three units as x increases four units anywhere along the line.

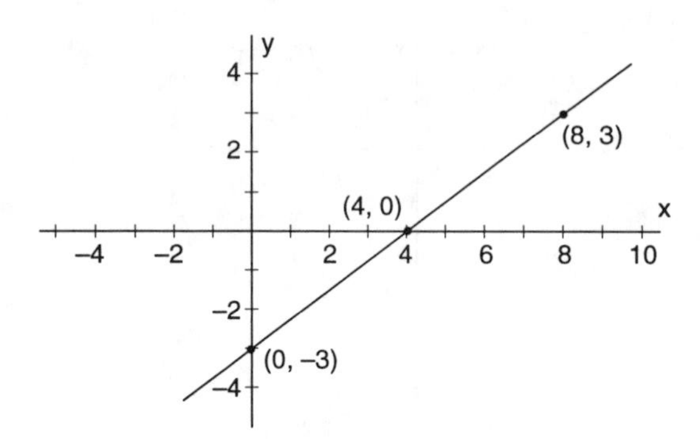

Figure 6.4

6.4 Systems of Linear Equations

A system of linear equations is a set of two or more linear equations as shown below:

$$\begin{cases} 2x+4y=11 \\ -5x+3y=5 \end{cases}$$

The set shown above is a system of linear equations with two variables, or unknowns.

There are several ways to solve systems of linear equations in two variables.

Method 1: Addition or subtraction—if necessary multiply the equations by numbers that will make the coefficients of one unknown in the resulting equations numerically equal. If the signs of equal coefficients are the same, subtract the equation, otherwise add.

The result is one equation with one unknown; we solve it and substitute the value into the other equations to find the unknown that we first eliminated.

Method 2: Substitution—find the value of one unknown in terms of the other; substitute this value in the other equation and solve.

Method 3: Graph—graph both equations. The point of intersection of the drawn lines is a simultaneous solution for the equations and its coordinates correspond to the answer that would be found analytically.

If the lines are parallel, they have no simultaneous solution.

Dependent equations are equations that represent the same line; therefore, every point on the line of a dependent equation represents a solution. Since there is an infinite number of points, there is an infinite number of simultaneous solutions, for example:

$$\begin{cases} 2x + y = 8 \\ 4x + 2y = 16 \end{cases}$$

The equations above are dependent; they represent the same line. All points that satisfy either of the equations are solutions of the system.

A system of linear equations is consistent if there is only one solution for the system.

A system of linear equations is inconsistent if it does not have any solutions.

6.4.1 Example of a Consistent System

Find the point of intersection of the graphs of the equations:

$$\begin{cases} x + y = 3 \\ 3x - 2y = 14 \end{cases}$$

Solution: To solve these linear equations, solve for y in terms of x. The equations will be in the form $y = mx + b$, where m is the slope and b is the intercept on the y-axis.

$$x + y = 3$$

$$y = 3 - x \qquad \text{Subtract } x \text{ from both sides.}$$

$$3x - 2y = 14 \qquad \text{Subtract } 3x \text{ from both sides.}$$

$$-2y = 14 - 3x \qquad \text{Divide by } -2.$$

$$y = -7 + \frac{3}{2}x$$

The graphs of the linear functions, $y = 3 - x$ and $y = -7 + \frac{3}{2}x$, can be determined by plotting only two points. For example, for $y = 3 - x$, let $x = 0$, then $y = 3$. Let $x = 1$, then $y = 2$. The two points on this first line are (0, 3) and (1, 2). For $y = -7 + \frac{3}{2}x$, let $x = 0$, then $y = -7$. Let $x = 1$, then $y = -5\frac{1}{2}$. The two points on this second line are (0, –7) and $\left(1,\ -5\frac{1}{2}\right)$. See Figure 6.5.

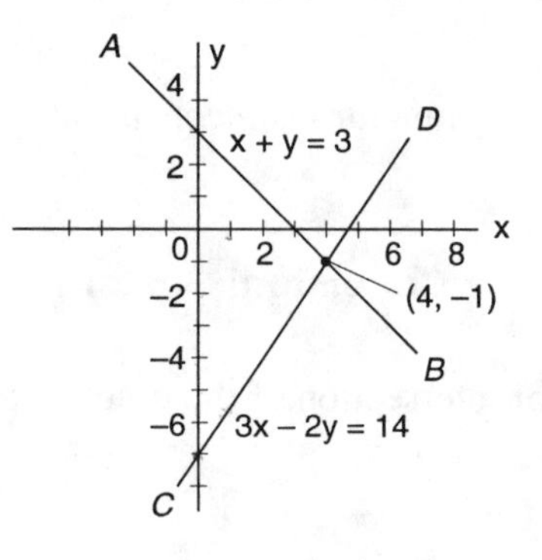

Figure 6.5

To find the point of intersection Figure 6.5 of

$$x + y = 3$$

and

$$3x - 2y = 14,$$

solve them algebraically. Multiply the first equation by 2. Add these two equations to eliminate the variable y.

$$\begin{array}{r} 2x + 2y = 6 \\ 3x - 2y = 14 \\ \hline 5x = 20 \end{array}$$

Solve for x to obtain $x = 4$. Substitute this into $y = 3 - x$ to get $y = 3 - 4 = -1$. P is $(4, -1)$. AB is the graph of the first equation, and CD is the graph of the second equation. The point of intersection P of the two graphs is the only point on both lines. The coordinates of P satisfy both equations and represent the desired solution of the problem. From the graph, P seems to be the point $(4, -1)$. These coordinates satisfy both equations, and hence are the exact coordinates of the point of intersection of the two lines.

To show that $(4, -1)$ satisfies both equations, substitute this point into both equations.

$$\begin{aligned} x + y &= 3 \\ 4 + (-1) &= 3 \\ 4 - 1 &= 3 \\ 3 &= 3 \end{aligned} \qquad \begin{aligned} 3x - 2y &= 14 \\ 3(4) - 2(-1) &= 14 \\ 12 + 2 &= 14 \\ 14 &= 14 \end{aligned}$$

6.4.2 Example of an Inconsistent System

Solve the equations $2x + 3y = 6$ and $4x + 6y = 7$ simultaneously (Figure 6.6).

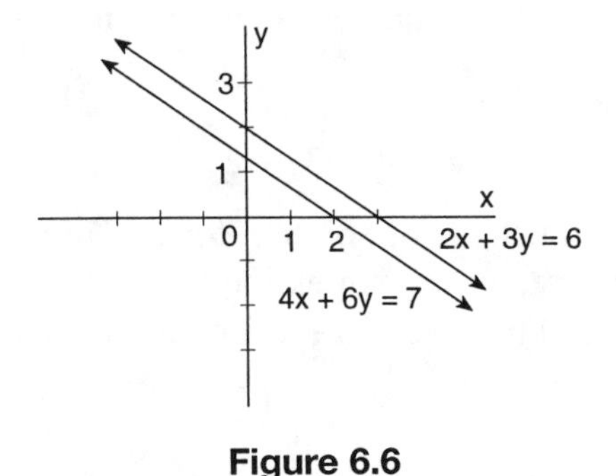

Figure 6.6

Solution: We have two equations in two unknowns,

$$2x + 3y = 6 \quad (1)$$

and

$$4x + 6y = 7 \quad (2)$$

There are several methods to solve this problem. We have chosen to multiply each equation by a different number so that when the two equations are added, one of the variables drops out. Thus,

multiplying Equation (1) by 2: $4x + 6y = 12$ (3)

multiplying Equation (2) by –1: $-4x - 6y = -7$ (4)

adding Equations (3) and (4): $0 = 5$

We obtain a peculiar result!

Actually, what we have shown in this case is that if there were a simultaneous solution to the given equations, then 0 would equal 5. But the conclusion is impossible; therefore, there can be no simultaneous solution, hence no point satisfies both equations.

The straight lines which are the graphs of these equations must be parallel if they never intersect, but are not identical, which can be seen from the graph of these equations (see Figure 6.6).

6.4.3 Example of a Dependent System

Solve the equations $2x + 3y = 6$ and $y = -(2x/3) + 2$ simultaneously.

Solution: We have two equations in two unknowns,

$$2x + 3y = 6 \quad (1)$$

$$y = -(2x/3) + 2 \quad (2)$$

There are several methods of solution for this problem. Since equation (2) already gives us an expression for y, we use the method of substitution. Substituting $-(2x/3) + 2$ for y in the first equation:

$$2x + 3\left(-\frac{2x}{3} + 2\right) = 6$$

Distributing, $$2x - 2x + 6 = 6$$

$$6 = 6$$

Apparently we have gotten nowhere! The result $6 = 6$ is true, but indicates no solution. Actually, our work shows that no matter what real number x is, if y is determined by the second equation then the first equation will always be satisfied.

The reason for this peculiarity may be seen if we take a closer look at the equation $y = -(2x/3) + 2$. It is equivalent to $3y = -2x + 6$, or $2x + 3y = 6$.

In other words, the two equations are equivalent. Any pair of values of x and y which satisfies one satisfies the other.

It is hardly necessary to verify that in this case the graphs of the given equations are identical lines, and that there are an infinite number of simultaneous solutions of these equations.

A system of three linear equations in three unknowns is solved by eliminating one unknown from any two of the three equations and solving them. After finding two unknowns, substitute them in any of the equations to find the third unknown.

Example: Solve the system

$$2x + 3y - 4z = -8 \qquad (1)$$

$$x + y - 2z = -5 \qquad (2)$$

$$7x - 2y + 5z = 4 \qquad (3)$$

Solution: We cannot eliminate any variable from any pair of equations by addition or subtraction. However, both x and z may be eliminated from Equations (1) and (2) by multiplying Equation (2) by -2. Then

$$2x + 3y - 4z = -8 \qquad (1)$$

$$-2x - 2y + 4z = 10 \qquad (4)$$

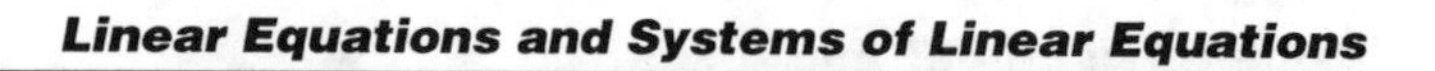

By addition, we have $y = 2$. Although we may now eliminate either x or z from another pair of equations, we can more conveniently substitute $y = 2$ in Equations (2) and (3) to get two equations in two variables. Thus, making the substitution $y = 2$ in Equations (2) and (3), we have

$$x - 2z = -7 \tag{5}$$

$$7x + 5z = 8 \tag{6}$$

Multiply Equation (5) by 5 and multiply Equation (6) by 2. Then add the two new equations. Then $x = -1$. Substitute x in either (5) or (6) to find z.

The solution of the system is $x = -1$, $y = 2$, and $z = 3$. Check by substitution.

A system of equations, as shown below, that has all constant terms $b_1, b_2, \dots b_n$ equal to zero is said to be a homogeneous system:

$$\begin{cases} a_{11}x_1 + a_{12}x_2 + \dots + a_{1n}x_m = b_1 \\ a_{21}x_1 + a_{22}x_2 + \dots + a_{2n}x_m = b_2 \\ \vdots \quad\quad \vdots \quad\quad\quad \vdots \quad\quad \vdots \\ a_{n1}x_1 + a_{n2}x_2 + \dots + a_{nn}x_m = b_n \end{cases}$$

A homogeneous system always has at least one solution which is called the trivial solution that is $x_1 = 0, x_2 = 0, \dots, x_m = 0$.

For any given homogeneous system of equations, in which the number of variables is greater than or equal to the number of equations, there are nontrivial solutions.

Two systems of linear equations are said to be equivalent if and only if they have the same solution set.

Problem Solving Examples:

Solve the equations $3x + 2y = 1$ and $5x - 3y = 8$ simultaneously.

We have two equations in two unknowns,

$$3x + 2y = 1 \quad (1)$$

and

$$5x - 3y = 8 \quad (2)$$

There are several methods to solve this problem. We have chosen to multiply each equation by a different number so that when the two equations are added, one of the variables drops out. Thus

multiplying the first by 3: $9x + 6y = 3$

and the second by 2: $10x - 6y = 16$

and add: $19x = 19$

$$x = 1$$

Substituting $x = 1$ in the first equation:

$$3(1) + 2y = 3 + 2y = 1$$

$$2y = -2$$

$$y = -1$$

(Alternatively, y might have been found by multiplying the first equation by 5, the second by –3, and adding.)

In this case, then, there is a unique solution: $x = 1$ and $y = -1$. This may be checked by replacing x by 1 and y by (–1) in each equation. In Equation (1):

$$3x + 2y = 1$$

$$3(1) + 2(-1) = 1$$

$$3 - 2 = 1$$

$$1 = 1$$

In Equation (2):

$$5x - 3y = 8$$
$$5(1) - 3(-1) = 8$$
$$5 - (-3) = 8$$
$$5 + 3 = 8$$
$$8 = 8$$

In other words, the lines whose equations are $3x + 2y = 1$ and $5x - 3y = 8$ meet in one and only one point: $(1, -1)$. This, again, may be checked graphically, as seen in the Figure 6.7.

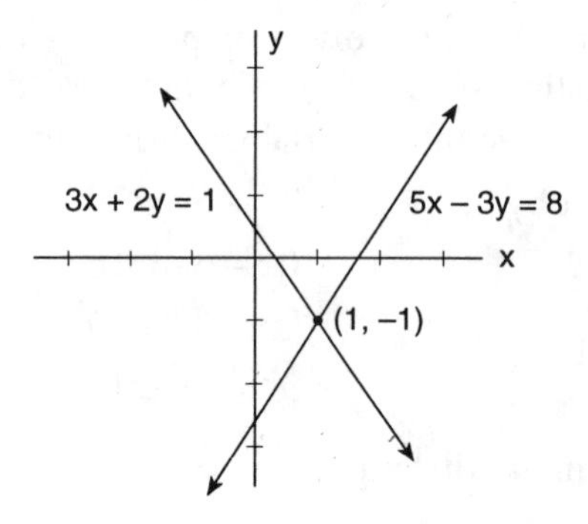

Figure 6.7

Solve for x and y.

$$x + 2y = 8 \quad (1)$$
$$3x + 4y = 20 \quad (2)$$

Solve Equation (1) for x in terms of :

$$x = 8 - 2y \quad (3)$$

Substitute $(8 - 2y)$ for x in (2):

$$3(8 - 2y) + 4y = 20 \quad (4)$$

Solve (4) for y as follows:

Distribute $\quad 24 - 6y + 4y = 20$

Combine like terms and then subtract 24 from both sides:

$$24 - 2y = 20$$
$$24 - 24 - 2y = 20 - 24$$
$$-2y = -4$$

Divide both sides by –2: $\quad y = 2$

Substitute 2 for y in Equation (1):

$$x + 2(2) = 8$$
$$x = 4$$

Thus, our solution is the point (4, 2).

Check: Substitute $x = 4$, $y = 2$ in Equations (1) and (2):

$$4 + 2(2) = 8$$
$$8 = 8$$
$$3(4) + 4(2) = 20$$
$$20 = 20$$

Quiz: Equations—Linear Equations

1. The solution set to the system

$$\begin{cases} 2x + 3y = 6 \\ y - 2 = -\dfrac{2x}{3} \end{cases}$$

is

(A) (0, 2).

(B) (0, 0).

(C) $\left(\frac{1}{3}, 5\right)$.

(D) (8, 3).

(E) Infinite number of solutions.

2. What is the solution to the pair of equations below?

$$\begin{cases} x - 3y = 1 \\ 2x + y = 2 \end{cases}$$

(A) $x = 1$ and $y = 0$

(B) $x = 2$ and $y = 0$

(C) $x = 3$ and $y = 1$

(D) $x = 0$ and $y = 1$

(E) $x = 0$ and $y = 2$

3. $4x - 2 = 10$

(A) $x = -1$

(B) $x = 2$

(C) $x = 3$

(D) $x = 4$

(E) $x = 6$

4. $7z + 1 - z = 2z - 7$

(A) $x = -2$

(B) $x = 0$

(C) $x = 1$

(D) $x = 2$

(E) $x = 3$

5. ${}^{1}/_{3}\, x + 3 = {}^{1}/_{2}\, x$

(A) $x = {}^{1}/_{2}$

(B) $x = 2$

(C) $x = 3^{3}/_{5}$

(D) $x = 6$

(E) $x = 18$

6. $0.4x + 1 = 0.7x - 2$

(A) $x = 0.1$
(B) $x = 2$
(C) $x = 5$
(D) $x = 10$
(E) $x = 12$

7. $4(3x + 2) - 11 = 3(3x - 2)$

(A) $x = -3$
(B) $x = -1$
(C) $x = 2$
(D) $x = 3$
(E) $x = 7$

8. $\begin{cases} 6x + 5y = -4 \\ 3x - 3y = 9 \end{cases}$

(A) $(1, -2)$
(B) $(1, 2)$
(C) $(2, -1)$
(D) $(-2, 1)$
(E) $(-1, 2)$

9. $\begin{cases} 4x + 3y = 9 \\ 2x - 2y = 8 \end{cases}$

(A) $(-3, 1)$
(B) $(1, -3)$
(C) $(3, 1)$
(D) $(3, -1)$
(E) $(-1, 3)$

10. $\begin{cases} x + y = 7 \\ x - y = -3 \end{cases}$

(A) $(5, 2)$
(B) $(-5, 2)$
(C) $(2, 5)$
(D) $(-2, 5)$
(E) $(2, -5)$

ANSWER KEY

1. (E)
2. (A)
3. (C)
4. (A)
5. (E)
6. (D)
7. (B)
8. (A)
9. (D)
10. (C)

CHAPTER 7

Inequalities

7.1 Inequality

An inequality is a statement that the value of one quantity or expression is greater than or less than that of another.

Example: $5 > 4$

The expression above means that the value of 5 is greater than the value of 4.

A conditional inequality is an inequality whose validity depends on the values of the variables in the sentence. That is, certain values of the variables will make the sentence true, and others will make it false. $3 - y > 3 + y$ is a conditional inequality for the set of real numbers, since it is true for any replacement less than zero and false for all others.

$x + 5 > x + 2$ is an absolute inequality for the set of real numbers, meaning that for any real valued x, the expression on the left is greater than the expression on the right.

$5y < 2y + y$ is inconsistent for the set of non-negative real numbers. For any x greater than 0, the sentence is always false. A sentence is inconsistent if it is always false when its variables assume allowable values.

The solution of a given inequality in one variable x consists of all values of x for which the inequality is true.

The graph of an inequality in one variable is represented by either a ray or a line segment on the real number line.

The endpoint is not a solution if the variable is strictly less than or greater than a particular value.

Example: $x > 2$

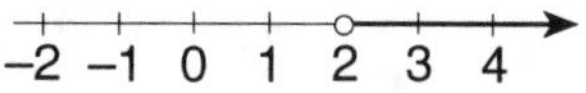

–2 –1 0 1 2 3 4

2 is not a solution and should be represented as shown.

The endpoint is a solution if the variable is either 1) less than or equal to or 2) greater than or equal to, a particular value.

Example: $5 > x \geq 2$

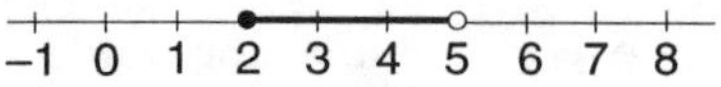

–1 0 1 2 3 4 5 6 7 8

In this case 2 is a solution and should be represented as shown.

7.2 Properties of Inequalities

If x and y are real numbers, then one and only one of the following statements is true.

$x > y$, $x = y$, or $x < y$.

This is the order property of real numbers.

If a, b, and c are real numbers:

A) If $a < b$ and $b < c$, then $a < c$.

B) If $a > b$ and $b > c$, then $a > c$.

This is the *transitive property of inequalities.*

If a, b, and c are real numbers and $a > b$, then $a + c > b + c$ and $a - c > b - c$. This is the *addition property of inequality.*

Two inequalities are said to have the same sense if their signs of inequality point in the same direction.

The *sense of an inequality* remains the same if both sides are multiplied or divided by the same positive real number.

Example: $4 > 3$

If we multiply both sides by 5, we will obtain:

$$4 \times 5 > 3 \times 5$$
$$20 > 15$$

The sense of the inequality does not change.

The sense of an inequality becomes opposite if each side is multiplied or divided by the same negative real number.

Example: $4 > 3$

If we multiply both sides by –5, we would obtain:

$$4 \times (-5) < 3 \times (-5)$$
$$-20 < -15$$

The sense of the inequality becomes opposite.

If $a > b$ and a, b, and n are positive real numbers, then:

$$a^n > b^n \text{ and } a^{-n} < b^{-n}$$

If $x > y$ and $q > p$ then $x + q > y + p$.

If $x > y > 0$ and $q > p > 0$, then $xq > yp$.

Inequalities that have the same solution set are called *equivalent inequalities.*

Problem Solving Examples:

Solve the inequality $2x + 5 > 9$.

$2x + 5 + (-5) > 9 + (-5)$	Add –5 to both sides
$2x + 0 > 9 + (-5)$	Additive inverse property
$2x > 9 + (-5)$	Additive identity property
$2x > 4$	Combine terms
$\frac{1}{2}(2x) > \frac{1}{2} \times 4$	Multiply both sides by $\frac{1}{2}$
$x > 2$	

The solution set is

$$X = \{x \mid 2x + 5 > 9\}$$
$$= \{x \mid x > 2\}$$

(that is all x, such that x is greater than 2).

Solve the inequality $4x + 3 < 6x + 8$.

In order to solve the inequality $4x + 3 < 6x + 8$, we must find all values of x which make it true. Thus, we wish to obtain x alone on one side of the inequality.

Add – 3 to both sides:

$$\begin{array}{rcl} 4x + 3 & < & 6x + 8 \\ -3 & & -3 \\ \hline 4x & < & 6x + 5 \end{array}$$

Add – $6x$ to both sides:

$$\begin{array}{rcl} 4x & < & 6x + 5 \\ -6x & & -6x \\ \hline -2x & < & 5 \end{array}$$

In order to obtain x alone, we must divide both sides by (– 2). Recall that dividing an inequality by a negative number reverses the inequality sign, hence

$$\frac{-2x}{-2} > \frac{5}{-2}$$

Cancelling $\frac{-2}{-2}$, we obtain, $x > -\frac{5}{2}$.

Thus, our solution is $\left\{x \mid x > -\frac{5}{2}\right\}\left(\right.$ the set of all x such that x is greater than $\left.-\frac{5}{2}\right)$.

7.3 Inequalities with Absolute Values

The solution set of $|a| < b$, $b > 0$ is $\{a \mid -b < a < b\}$.

The solution set of $|a| > b$ is $\{a \mid a > b \text{ or } a < -b\}$.

Problem Solving Examples:

Express the inequality $|x| < 3$ without using absolute value signs.

According to the law of absolute values which states that $|a| < b$ is equivalent to $-b < a < b$, where b is any positive number, $|x| < 3$ is equivalent to $-3 < x < 3$.

Solve the inequality $| 5 - 2x | > 3$.

Use the law of absolute values, which states $|a| > b$ is $\{a \mid a > b$ or $a < -b\}$. Therefore:

$$5 - 2x > 3 \qquad \text{or} \qquad 5 - 2x < -3$$

$$-2x > -2 \qquad\qquad -2x < -8$$

$$x < 1 \qquad\qquad x > 4$$

Therefore, the above inequality holds when $x < 1$, and when $x > 4$.

7.4 Inequalities in Two Variables

An inequality of the form $ax + by < c$ is a linear inequality in two variables. The equation for the boundary of the solution set is given by $ax + by = c$.

To graph a linear inequality, first graph the boundary.

Next, choose any point off the boundary and substitute its coordinates into the original inequality. If the resulting statement is true, the graph lies on the same side of the boundary as the test point. A false statement indicates that the solution set lies on the other side of the boundary.

Problem Solving Examples:

Solve $2x - 3y \geq 6$.

The statement $2x - 3y \geq 6$ means $2x - 3y$ is greater than or equal to 6. Symbolically, we have $2x - 3y > 6$ or $2x - 3y = 6$. Consider the corresponding equality and graph $2x - 3y = 6$. To find the x-intercept, set $y = 0$

$$2x - 3y = 6$$

$$2x - 3(0) = 6$$

$$2x = 6$$

$$x = 3$$

(3, 0) is the x-intercept.

To find the y-intercept, set $x = 0$

$$2x - 3y = 6$$

$$2(0) - 3y = 6$$

$$-3y = 6$$
$$y = -2$$

(0, –2) is the *y*-intercept.

A line is determined by two points. Therefore draw a straight line through the two intercepts (3, 0) and (0, –2). Since the inequality is mixed, a solid line is drawn through the intercepts. This line represents the part of the statement $2x - 3y = 6$.

We must now determine the region for which the inequality $2x - 3y > 6$ holds.

Choose two points to decide on which side of the line the region $x - 3y > 6$ lies. We shall try the points (0, 0) and (5, 1).

For (0, 0)	For (5, 1)
$2x - 3y > 6$	$2x - 3y > 6$
$2(0) - 3(0) > 6$	$2(5) - 3(1) > 6$
$0 - 0 > 6$	$10 - 3 > 6$
$0 > 6$	$7 > 6$
False	True

The inequality, $2x - 3y > 6$, holds true for the point (5, 1). We shade this region of the *xy*-plane. That is, the area lying below the line $2x - 3y = 6$ and containing (5, 1), see Figure 7.1.

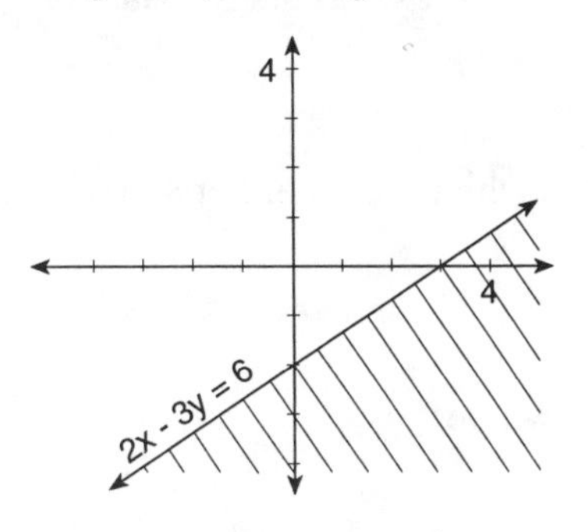

Figure 7.1

Therefore, the solution contains the solid line, $2x - 3y = 6$, and the part of the plane below this line for which the statement $2x - 3y > 6$ holds.

Solve the inequality $x + 2y \geq 6$ for y in terms of x and draw its graph.

A To solve for y in terms of x, obtain y alone on one side of the inequality and x and any constants on the other. Subtracting x from both sides of

$$x + 2y \geq 6 \text{ gives } 2y \geq 6 - x$$

Divide the equation by 2

$$y \geq 3 - \frac{1}{2}x$$

The points in the $x - y$ plane which will satisfy this equation are those satisfying

$$y > 3 - \frac{1}{2}x \text{ and } y = 3 - \frac{1}{2}x$$

Consider the case,

$$y = 3 - \frac{1}{2}x$$

which is a graph of the solid straight line with y-intercept (0, 3) and slope $-\frac{1}{2}$. (See Figure 7.2.)

We must also find those points that satisfy

$$y > 3 - \frac{1}{2}x.$$

Choose two points which lie on either side of the line

$$y = 3 - \frac{1}{2}x$$

to find the region where

$$y > 3 - \frac{1}{2}x.$$

We shall choose (3, 3) and (3, – 3) (See Figure 7.2).

For (3, 3)

$$y > 3 - \frac{1}{2}x$$

$$3 > 3 - \frac{1}{2}(3)$$

$$3 > \frac{3}{2}$$

(3, 3) satisfies the inequality.

For (3, – 3)

$$y > 3 - \frac{1}{2}x$$

$$-3 > 3 - \frac{1}{2}(3)$$

$$-3 \not> \frac{3}{2}$$

(3, – 3) does not satisfy the inequality.

Thus, all the points in the region where (3, 3) lies satisfy $y > 3 - \frac{1}{2}x$. That is all those points above the line $y = 3 - \frac{1}{2}x$ satisfy $y > 3 - \frac{1}{2}x$ and lie in the shaded area (see Figure 7.2).

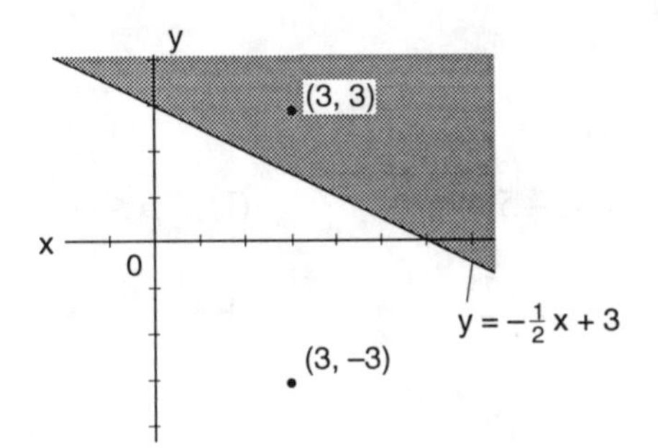

Figure 7.2

Notice that when the line is in slope-intercept form, greater than means above the given line while less than means below the given line.

Consequently, the graphical solution of $y \geq -\frac{1}{2}x + 3$ are those points that lie on the solid line $y = -\frac{1}{2}x + 3$ and those points in the shaded area $y > -\frac{1}{2}x + 3$.

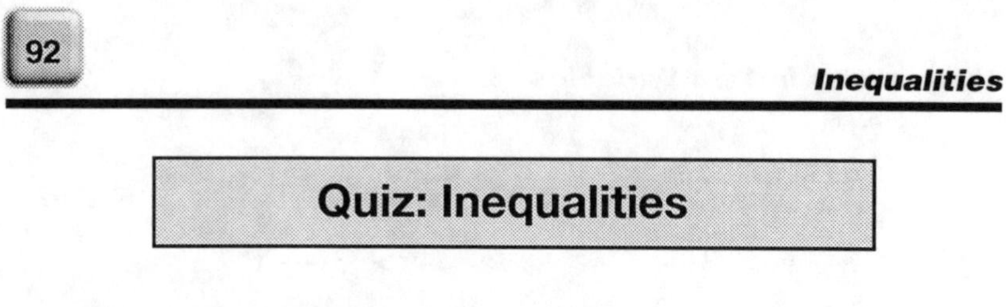

Quiz: Inequalities

Find the solution set for each inequality.

1. $3m + 2 < 7$

   (A) $m \geq 5/3$
   (B) $m \leq 2$
   (C) $m < 2$
   (D) $m > 2$
   (E) $m < 5/3$

2. $1/2\, x - 3 \leq 1$

   (A) $-4 \leq x \leq 8$
   (B) $x \geq -8$
   (C) $x \leq 8$
   (D) $2 \leq x \leq 8$
   (E) $x \geq 8$

3. $-3p + 1 \geq 16$

   (A) $p \geq -5$
   (B) $p \geq \frac{-17}{3}$
   (C) $p \leq \frac{-17}{3}$
   (D) $p \leq -5$
   (E) $p \geq 5$

4. $-6 < 2/3\, r + 6 \leq 2$

   (A) $-6 < r \leq -3$
   (B) $-18 < r \leq -6$
   (C) $r \geq -6$
   (D) $-2 < r \leq -4/3$
   (E) $r \leq -6$

5. $0 < 2 - y < 6$

(A) $-4 < y < 2$
(B) $-4 < y < 0$
(C) $-4 < y < -2$
(D) $-2 < y < 4$
(E) $0 < y < 4$

6. What is the range of values for which $|6x - 5| \leq 8$ is satisfied?

(A) $-\frac{1}{2} \leq x \leq \frac{1}{2}$
(B) $0 \leq x \leq \frac{5}{6}$
(C) $-1 \leq x \leq \frac{1}{2}$
(D) $-\frac{1}{2} \leq x \leq \frac{13}{6}$
(E) $-\frac{1}{2} \leq x \leq \frac{1}{3}$

7. If $5 < a < 8$ and $6 < b < 9$, then

(A) $45 < ab < 48$.
(B) $30 < ab < 45$.
(C) $30 < ab < 72$.
(D) $54 < ab < 72$.
(E) $5 < ab < 72$.

8. What is the solution set of $|3x - 9| < 5$?

(A) $\frac{4}{3} < x < \frac{8}{3}$
(B) $-\frac{4}{3} < x < \frac{7}{3}$
(C) $-\frac{4}{3} < x < \frac{8}{3}$
(D) $\frac{4}{3} < x < \frac{13}{3}$
(E) $\frac{4}{3} < x < \frac{14}{3}$

9. If $-9 < x < -4$ and $-12 < y < -6$, then

(A) $0 < xy < 12$.

(B) $108 < xy < 112$.

(C) $24 < xy < 108$.

(D) $10 < xy < 24$.

(E) $4 < xy < 12$.

10. If a function is defined as $|2 - 5x| < 3$, then the interval which does not contain any solution for x is

(A) $0 < x < 1$.

(B) $0 < x < 2$.

(C) $-\frac{1}{25} < x < 0$.

(D) $-\frac{3}{5} < x < -\frac{1}{2}$.

(E) $-1 < x < 1$.

ANSWER KEY

1. (E)
2. (C)
3. (D)
4. (B)
5. (A)
6. (D)
7. (C)
8. (E)
9. (C)
10. (D)

CHAPTER 8

Relations and Functions

8.1 Relations and Functions

A relation is any set of ordered pairs. The set of all first members of the ordered pairs is called the domain of the relation and the set of all second members of the ordered pairs is called the range of the relation.

Example: Find the relation defined by $y^2 = 25 - x^2$ where the domain $D = \{0, 3, 4, 5\}$.

Solution: x takes on the values 0, 3, 4, and 5. Replacing x by these values in the equation $y^2 = 25 - x^2$, we obtain the corresponding values of y (see Table 8.1).

Hence the relation defined by $y^2 = 25 - x^2$ where x belongs to $D = \{0, 3, 4, 5\}$ is

$$\{(0, 5), (0, -5), (3, 4), (3, -4), (4, 3), (4, -3), (5, 0)\}.$$

The domain of the relation is (0, 3, 4, 5). The range of the relation is (5, –5, 4, – 4, 3, –3, 0).

A function with domain X and range Y could be given by Table 8.1:

Table 8.1

x	$y^2 = 25 - x^2$	y
0	$y^2 = 25 - 0$ $y^2 = 25$ $y = \sqrt{25}$ $y = \pm 5$	± 5
3	$y^2 = 25 - 3^2$ $y^2 = 25 - 9$ $y^2 = 16$ $y = \sqrt{16}$ $y = \pm 4$	± 4
4	$y^2 = 25 - 4^2$ $y^2 = 25 - 16$ $y^2 = 9$ $y = \sqrt{9}$ $y = \pm 3$	± 3
5	$y^2 = 25 - 5^2$ $y^2 = 25 - 25$ $y^2 = 0$ $y = 0$	0

$\{(1, 2), (2, 2), (3, 4), (4, 4), (5, 6), (6, 6), (7, 8), (8, 8)\}$

A function is a relation in which no two ordered pairs have the same first member. For example:

$X = \{1, 2, 3, 4, 5, 6, 7, 8\}$ and $Y = \{2, 4, 6, 8\}$

You can see above that every member of the domain is paired with one and only one member of the range. Then this relation is called a function and is represented by $y = f(x)$, where $x \in X$ and $y \in Y$. If f is a function that takes an element $x \in X$ and sends it to an element $y \in Y$, f is said to map x into y. We write this as $f : x \rightarrow y$. For this reason, a function is also called a mapping.

Given $f{:}x \rightarrow y$, we can also say that y is a function of x, denoted $f(x) = y$, "f of x equals y." In this function, y is called the dependent variable, since it derives its value from x. By the same reasoning, x is called the independent variable.

Another way of checking if a relation is a function is the vertical line test: if there does not exist any vertical line which crosses the graph of a relation in more than one place, then the relation is a function. If the domain of a relation or a function is not specified, it is assumed to be all real numbers.

Problem Solving Examples:

Find the domain D and the range R of the function $\left(x, \frac{x}{|x|}\right)$.

Note that the y-value of any coordinate pair (x, y) is $\frac{x}{|x|}$. We can replace x in the formula $\frac{x}{|x|}$ with any number except 0, since the denominator, $|x|$, cannot equal 0 (i.e., $|x| \neq 0$). This is because division by 0 is undefined. Therefore, the domain D is the set of all real numbers except 0. If x is negative, i.e., $x < 0$, then $|x| = -x$ by definition. Hence, if x is negative, then $\frac{x}{|x|} = \frac{x}{-x} = -1$. If x is positive, i.e. $x > 0$, then $|x| = x$ by definition. Hence, if x is positive, then $\frac{x}{|x|} = \frac{x}{x} = 1$. (The case where $x = 0$ has already been found to be undefined.) Thus, there are only two numbers -1 and 1 in the range R of the function; that is, $R = \{-1, 1\}$.

Find the set of ordered pairs $\{(x, y)\}$ if $y = x^2 - 2x - 3$ and $D = \{x \mid x \text{ is an integer and } 1 \leq x \leq 4\}$.

We first note that $D = \{1, 2, 3, 4\}$. Substituting these values of x in the equation

$y = x^2 - 2x - 3,$

we find the corresponding y values. Thus,

for $x = 1$, $y = 1^2 - 2(1) - 3 = 1 - 2 - 3 = -4$

for $x = 2$, $y = 2^2 - 2(2) - 3 = 4 - 4 - 3 = -3$

for $x = 3$, $y = 3^2 - 2(3) - 3 = 9 - 6 - 3 = 0$

for $x = 4$, $y = 4^2 - 2(4) - 3 = 16 - 8 - 3 = 5$

Hence, $\{(x, y)\} = \{(1, -4), (2, -3), (3, 0), (4, 5)\}$.

8.2 Properties of Relations

A relation R from set A to set B is a subset of the Cartesian Product $A \times B$ written aRb with $a \in A$ and $b \in B$.

Let R be a relation from a set S to itself. Then

A) R is said to be reflexive if and only if sRs for every $s \in S$.

B) R is said to be symmetric if $s_iRs_j \Rightarrow s_jRs_i$ where $s_i, s_j \in S$.

C) R is said to be transitive if s_iRs_j and s_jRs_k implies s_iRs_k.

D) R is said to be antisymmetric if s_1Rs_2 and s_2Rs_1 implies $s_1 = s_2$.

A relation R on $S \times S$ is called an equivalence relation if R is reflexive, symmetric, and transitive.

8.3 Properties of Functions

If f and g are two functions with a common domain, then the sum of f and g, written $f + g$, is defined by:

$$(f + g)(x) = f(x) + g(x)$$

The difference of f and g is defined by:

$$(f - g)(x) = f(x) - g(x)$$

The quotient of f and g is defined by:

$$\left(\frac{f}{g}\right)(x) = \frac{f(x)}{g(x)}, \text{ where } g(x) \neq 0$$

Example: Let $f(x) = 2x^2$ with domain $D_f = R$ (or, alternatively, C) and $g(x) = x - 5$ with $D_g = R$ (or C). Find A) $f + g$, B) $f - g$, C) fg, and D) $\frac{f}{g}$.

A) $f + g$ has domain R (or C) and

$$(f + g)(x) = f(x) + g(x) = 2x^2 + x - 5$$

for each number x. For example, $(f + g)(1) = f(1) + g(1) = 2(1)^2 + 1 - 5 = 2 - 4 = -2$.

B) $f - g$ has domain R (or C) and

$$(f - g)(x) = f(x) - g(x) = 2x^2 - (x - 5) = 2x^2 - x + 5$$

for each number x. For example, $(f - g)(1) = f(1) - g(1) = 2(1)^2 - 1 + 5 = 2 + 4 = 6$.

C) fg has domain R (or C) and

$$(fg)(x) = f(x) \bullet g(x) = 2x^2 \bullet (x - 5) = 2x^3 - 10x^2$$

for each number x. In particular, $(fg)(1) = 2(1)^3 - 10(1)^2 = 2 - 10 = -8$.

D) $\frac{f}{g}$ has domain R (or C) excluding the number $x = 5$ (when $x = 5$, $g(x) = 0$ and division by zero is undefined) and

$$\left(\frac{f}{g}\right)(x) = \frac{f(x)}{g(x)} = \frac{2x^2}{x - 5}$$

for each number $x \neq 5$. In particular, $\left(\frac{f}{g}\right)(1) = \frac{2(1)^2}{1 - 5} = \frac{2}{-4} = -\frac{1}{2}$.

If f is a function, then the inverse of f, written f^{-1} is such that:

$$(x, y) \in f \Leftrightarrow (y, x) \in f^{-1}$$

The graph of f^{-1} can be obtained from the graph of f by simply reflecting the graph of f across the line $y = x$. The graphs of f and f^{-1} are symmetrical about the line $y = x$.

The inverse of a function is not necessarily a function.

Example: Show that the inverse of the function $y = x^2 + 4x - 5$ is not a function.

Solution: Given the function f such that no two of its ordered pairs have the same first element, the inverse function f^{-1} is the set of ordered pairs obtained from f by interchanging in each ordered pair the first and second elements. Thus, the inverse of the function

$$y = x^2 + 4x - 5 \text{ is } x = y^2 + 4y - 5.$$

The given function has more than one first component corresponding to a given second component. For example, if $y = 0$, then $x = -5$ or 1. If the elements $(-5, 0)$ and $(1, 0)$ are reversed, we have $(0, -5)$ and $(0, 1)$ as elements of the inverse. Since the first component 0 has more than one second component, the inverse is not a function (a function can have only one y value corresponding to each x value).

A function $f : A \rightarrow B$ is said to be one-to-one or injective if distinct elements in the domain A have distinct images, i.e. if $f(x) = f(y)$ implies $x = y$. For an example: $y = f(x) = x^3$ defined over the domain $\{x \in R \mid x \geq 0\}$ is an injection or an injective function.

A function $f : A \rightarrow B$ is said to be a surjective or an onto function if each element of B is the image of some element of A, i.e., $f(A) = B$. For instance, $y = x \sin x$ is a surjection or a surjective function.

A function $f : A \rightarrow B$ is said to be bijective or a bijection if f is both injective and surjective. f is also called a one-to-one, onto correspondence between A and B. An example of such function would be $y = x$.

Problem Solving Example:

If $f(x) = x^2 - x - 3$, $g(x) = (x^2 - 1)/(x + 2)$, and $h(x) = f(x) + g(x)$, find $h(2)$.

$h(x) = f(x) + g(x)$, and we are told that $f(x) = x^2 - x - 3$ and $g(x) = x^2 - 1/x + 2$; thus, $h(x) = (x^2 - x - 3) + (x^2 - 1)/(x + 2)$.

To find $h(2)$, we replace x by 2 in the above formula for $h(x)$,

$$\begin{aligned} h(2) &= [(2)^2 - 2 - 3] + \left(\frac{2^2+1}{2+2}\right) \\ &= (4 - 2 - 3) + \left(\frac{4-1}{4}\right) \\ &= (-1) + \left(\frac{3}{4}\right) \\ &= -\frac{4}{4} + \frac{3}{4} \\ &= -\frac{1}{4} \end{aligned}$$

Thus, $h(2) = -\frac{1}{4}$.

Quiz: Relations and Functions

1. If $f(x) = 7x^2 + 3$ and $g(x) = 2x - 9$, then $g(f(2)) =$

   (A) 28. (D) 19.

   (B) 0. (E) 53.

   (C) 31.

2. A function is defined as $f(x) = x^2 + 2$. What is the numerical value of $3f(0) + f(-1)f(2)$?

(A) 6
(B) 24
(C) 18
(D) 4
(E) 36

3. If $f(x) = 3x + 2$ and $g(f(x)) = x$, then $g(x) =$

(A) $\frac{x-2}{3}$.
(B) $\frac{x}{3} - 2$.
(C) $3x$.
(D) $3x - 2$.
(E) $4x + 9$.

4. If $f(x) = 3x^2 - x + 5$, then $f(3) =$

(A) 15.
(B) 17.
(C) 23.
(D) 27.
(E) 29.

5. If $f(x) = \sqrt{x+1}$ and $g(x) = x^3 + 2$, find $f(g(2))$.

(A) 3.32
(B) 2.9
(C) 7.1
(D) 10.9
(E) 4.2

6. If $f(x) = 3x - 5$ and $g(f(x)) = x$, then $g(x) =$

(A) $\dfrac{x-5}{3}$.

(B) $\dfrac{x+5}{3}$.

(C) $\dfrac{2x+5}{3}$.

(D) $\dfrac{x+5}{4}$.

(E) $\dfrac{5-x}{3}$.

7. If $f(x) = 2^x + 4$ and $g(x) = \dfrac{1}{x}$, then $f(g(f(x)))$ is

(A) $\dfrac{1}{2^x+4}$.

(B) $2^{1/(2x+4)}+4$.

(C) $\dfrac{1}{2^{2x+4}}$.

(D) $\dfrac{1}{2^{2x}+4}+4$.

(E) $\dfrac{1}{2^{2x}}+4$.

8. Consider the function

$$f(x) = x^3 + 3x - k.$$

If $f(2) = 10$, then $k =$

(A) 4.

(B) −4.

(C) 0.

(D) 18.

(E) 14.

9. If $f(x) = \sqrt{x}$, $g(x) = \dfrac{x-1}{4}$, and $h(x) = x^2$, what is $f(g(h(4)))$?

(A) 2
(B) 2.06
(C) 2.24
(D) 1.94
(E) 3.75

10. If $f(x) = 2x + 1$ and $g(x) = 3x - 5$, then $f(g(2)) =$

(A) 0.
(B) 1.
(C) 2.
(D) 3.
(E) 5.

ANSWER KEY

1. (E)
2. (B)
3. (A)
4. (E)
5. (A)
6. (B)
7. (B)
8. (A)
9. (D)
10. (D)

CHAPTER 9

Quadratic Equations

9.1 Quadratic Equations

A second degree equation in x of the type $ax^2 + bx + c = 0$, $a \neq 0$, a, b, and c are real numbers, is called a quadratic equation.

To solve a quadratic equation is to find values of x which satisfy $ax^2 + bx + c = 0$. These values of x are called solutions, or roots, of the equation.

A quadratic equation has a maximum of two roots. Methods of solving quadratic equations:

A) Direct solution: Given $x^2 - 9 = 0$.

We can solve directly by isolating the variable x:

$$x^2 = 9$$

$$x = \pm 3$$

B) Factoring: Given a quadratic equation $ax^2 + bx + c = 0$, a, b, $c \neq 0$, to factor means to express it as the product $a(x - r_1)(x - r_2) = 0$, where r_1 and r_2 are the two roots.

Some helpful hints to remember are:

(a) $r_1 + r_2 = -\frac{b}{a}$.

(b) $r_1 + r_2 = \frac{c}{a}$.

Given $x^2 - 5x + 4 = 0$.

Since $r_1 + r_2 = \frac{-b}{a} = \frac{-(-5)}{1} = 5$, so the possible solutions are (3, 2), (4, 1), and (5, 0). Also $r_1 r_2 = \frac{c}{a} = \frac{4}{1} = 4$; this equation is satisfied only by the second pair, so $r_1 = 4$, $r_2 = 1$ and the factored form is $(x - 4)(x - 1) = 0$.

If the coefficient of x^2 is not 1, it is necessary to divide the equation by this coefficient and then factor.

Given $2x^2 - 12x + 16 = 0$.

Dividing by 2, we obtain:

$$x^2 - 6x + 8 = 0$$

Since $r_1 + r_2 = \frac{-b}{a} = 6$, the possible solutions are (6, 0), (5, 1), (4, 2), (3, 3). Also $r_1\ r_2 = 8$, so the only possible answer is (4, 2) and the expression $x^2 - 6x + 8 = 0$ can be factored as $(x - 4)(x - 2)$.

C) Completing the square: If it is difficult to factor the quadratic equation using the previous method, we can complete the square.

Given $x^2 - 12x + 8 = 0$.

We know that the two roots added up should be 12 because $r_1 + r_2 = \frac{-b}{a} = \frac{-(-12)}{1} = 12$. The possible roots are (12, 0), (11, 1), (10, 2), (9, 3), (8, 4), (7, 5), (6, 6).

But none of these satisfy $r_1\ r_2 = 8$, so we cannot use B).

To complete the square it is necessary to isolate the constant term,

$$x^2 - 12x = -8.$$

Then take $\frac{1}{2}$ coefficient of the x term, square it and add to both sides

$$x^2 - 12x + \left(\frac{-12}{2}\right)^2 = -8 + \left(\frac{-12}{2}\right)^2$$

$$x^2 - 12x + 36 = -8 + 36 = 28$$

Now we can use the previous method to factor the left side: $r_1 + r_2 = 12$, $r_1 r_2 = 36$ is satisfied by the pair (6, 6), so we have:

$$(x - 6)(x - 6) = (x - 6)^2 = 28.$$

Now take the square root of both sides and solve for x. Remember when taking a square root that the solution can be positive or negative.

$$(x - 6) = \pm\sqrt{28} = \pm 2\sqrt{7}$$

$$x = \pm 2\sqrt{7} + 6$$

So the roots are: $x = 2\sqrt{7} + 6,\ x = -2\sqrt{7} + 6.$

Problem Solving Example:

Solve $2x^2 + 8x + 4 = 0$ by completing the square.

Divide both members by 2, the coefficient of x^2.
$x^2 + 4x + 2 = 0$

Subtract the constant term, 2, from both members.

$$x^2 + 4x = -2$$

Add to each member the square of one-half the coefficient of the x-term.

$$x^2 + 4x + 4 = -2 + 4$$

Factor

$$(x + 2)(x + 2) = (x + 2)^2 = 2$$

Set the square root of the left member (a perfect square) equal to $\pm$ the square root of the right member and solve for x.

$$x + 2 = \sqrt{2} \qquad \text{or} \qquad x + 2 = -\sqrt{2}$$

The roots are $\sqrt{2} - 2$ and $-\sqrt{2} - 2$. Check each solution.

$$\begin{aligned} 2\left(\sqrt{2}-2\right)^2 + 8\left(\sqrt{2}-2\right)+4 &= 2\left(2-4\sqrt{2}+4\right)+8\sqrt{2}-16+4 \\ &= 4-8\sqrt{2}+8+8\sqrt{2}-16+4 \\ &= 0 \end{aligned}$$

$$\begin{aligned} 2\left(-\sqrt{2}-2\right)^2 + 8\left(-\sqrt{2}-2\right)+4 &= 2\left(2+4\sqrt{2}+4\right)-8\sqrt{2}-16+4 \\ &= 4+8\sqrt{2}+8-8\sqrt{2}-16+4 \\ &= 0 \end{aligned}$$

9.2 Quadratic Formula

Consider the polynomial:

$ax^2 + bx + c = 0$, where $a \neq 0$.

The roots of this equation can be determined in terms of the coefficients a, b, and c as shown below:

$$x = \frac{-b \pm \sqrt{b^2 - 4ac}}{2a}$$

where $(b^2 - 4ac)$ is called the discriminant of the quadratic equation.

Note that if the discriminant is less than zero ($b^2 - 4ac < 0$), the roots are complex numbers, since the discriminant appears under a radical and square roots of negatives are complex numbers, and a real number added to an imaginary number yields a complex number.

If the discriminant is equal to zero ($b^2 - 4ac = 0$), the result is one real root.

If the discriminant is greater than zero ($b^2 - 4ac > 0$), then the roots are real and unequal. Further, the roots are rational if and only if a and b are rational and ($b^2 - 4ac$) is a perfect square, otherwise the roots are irrational.

Example: Compute the value of the discriminant and then determine the nature of the roots of each of the following four equations:

A) $4x^2 - 12x + 9 = 0$

B) $3x^2 - 7x - 6 = 0$

C) $5x^2 + 2x - 9 = 0$

D) $x^2 + 3x + 5 = 0$

A) $4x^2 - 12x + 9 = 0$,

Here a, b, and c are integers,

$$a = 4, b = -12, \text{ and } c = 9.$$

Therefore,

$$b^2 - 4ac = (-12)^2 - 4(4)(9) = 144 - 144 = 0.$$

Since the discriminant is 0, the roots are rational and equal.

B) $3x^2 - 7x - 6 = 0$

Here a, b, and c are integers,

$$a = 3, \ b = -7, \ \text{and } c = -6.$$

Therefore,

$$b^2 - 4ac = (-7)^2 - 4(3)(-6) = 49 + 72 = 121 = 11^2.$$

Since the discriminant is a perfect square, the roots are rational and unequal.

C) $5x^2 + 2x - 9 = 0$

Here a, b, and c are integers,

$a = 5, b = 2$, and $c = -9$.

Therefore,

$$b^2 - 4ac = 2^2 - 4(5)(-9) = 4 + 180 = 184.$$

Since the discriminant is greater than zero, but not a perfect square, the roots are irrational and unequal.

D) $x^2 + 3x + 5 = 0$

Here a, b, and c are integers,

$a = 1, b = 3$, and $c = 5$.

Therefore,

$$b^2 - 4ac = 3^2 - 4(1)(5) = 9 - 20 = -11.$$

Since the discriminant is negative, the roots are imaginary.

Example: Find the equation whose roots are $\frac{\alpha}{\beta}, \frac{\beta}{\alpha}$.

Solution: The roots of the equation are $x = \frac{\alpha}{\beta}$ and $x = \frac{\beta}{\alpha}$. Subtract $\frac{\alpha}{\beta}$ from both sides of the first equation:

$$x - \frac{\alpha}{\beta} = \frac{\alpha}{\beta} - \frac{\alpha}{\beta} = 0$$

or

$$x - \frac{\alpha}{\beta} = 0$$

Subtract $\frac{\beta}{\alpha}$ from both sides of the second equation:

$$x - \frac{\beta}{\alpha} = \frac{\beta}{\alpha} - \frac{\beta}{\alpha} = 0$$

or

$$x - \frac{\beta}{\alpha} = 0$$

Therefore:

$$\left(x-\frac{\alpha}{\beta}\right)\left(x-\frac{\beta}{\alpha}\right)=(0)(0),$$

or

$$\left(x-\frac{\alpha}{\beta}\right)\left(x-\frac{\beta}{\alpha}\right)=0. \qquad (1)$$

Equation (1) is of the form:

$$(x-c)(x-d)=0, \text{ or}$$

$$x^2-cx-dx+cd=0, \text{ or}$$

$$x^2-(c+d)x+cd=0. \qquad (2)$$

Note that c corresponds to the root $\frac{\alpha}{\beta}$ and d corresponds to the root $\frac{\beta}{\alpha}$. The sum of the roots is:

$$c+d=\frac{\alpha}{\beta}+\frac{\beta}{\alpha}=\frac{\alpha(\alpha)}{\alpha(\beta)}+\frac{\beta(\beta)}{\beta(\alpha)}=\frac{\alpha^2}{\alpha\beta}+\frac{\beta^2}{\alpha\beta}$$

$$=\frac{\alpha^2+\beta^2}{\alpha\beta}$$

The product of the roots is:

$$c\times d=\frac{\alpha}{\beta}\times\frac{\beta}{\alpha}=\frac{\alpha\beta}{\beta\alpha}=\frac{\alpha\beta}{\alpha\beta}=1$$

Using the form of Equation (2):

$$\left(x-\frac{\alpha}{\beta}\right)\left(x-\frac{\beta}{\alpha}\right)=x^2-\left(\frac{\alpha^2+\beta^2}{\alpha\beta}\right)x+1=0.$$

Hence,

$$x^2 - \left(\frac{\alpha^2 + \beta^2}{\alpha\beta}\right)x + 1 = 0. \tag{3}$$

Multiply both sides of Equation (3) by $\alpha\beta$

$$\alpha\beta\left[x^2 - \left(\frac{\alpha^2 + \beta^2}{\alpha\beta}\right)x + 1\right] = \alpha\beta(0)$$

Distributing,

$$\alpha\beta x^2 - (\alpha^2 + \beta^2)x + \alpha\beta = 0,$$

which is the equation whose roots are $\frac{\alpha}{\beta}, \frac{\beta}{\alpha}$.

9.2.1 Radical Equation

An equation that has one or more unknowns under a radical is called a radical equation.

To solve a radical equation, isolate the radical term on one side of the equation and move all the other terms to the other side. Then both members of the equation are raised to a power equal to the index of the isolated radical.

After solving the resulting equation, the roots obtained must be checked, since this method often introduces extraneous roots.

These introduced roots must be excluded if they are not solutions.

Given $\sqrt{x^2 + 2} + 6x = x - 4$

$$\sqrt{x^2 + 2} = x - 4 - 6x = -5x - 4$$

$$\left(\sqrt{x^2 + 2}\right)^2 = (-(5x + 4))^2$$

$$x^2 + 2 = (5x + 4)^2$$

$$x^2 + 2 = 25x^2 + 40x + 16$$

$$24x^2 + 40x + 14 = 0$$

Applying the quadratic formula, we obtain:

$$x = \frac{-40 \pm \sqrt{1600 - 4(24)(14)}}{2(24)} = \frac{-40 \pm 16}{48}$$

$$x_1 = \frac{-7}{6}, \quad x_2 = \frac{-1}{2}$$

Checking roots:

$$\sqrt{\left(\frac{-7}{6}\right)^2 + 2} + 6\left(\frac{-7}{6}\right) \stackrel{?}{=} \left(\frac{-7}{6}\right) - 4$$

$$\frac{11}{6} - 7 \stackrel{?}{=} \frac{-31}{6}$$

$$\frac{-31}{6} = \frac{-31}{6}$$

$$\sqrt{\left(\frac{-1}{2}\right)^2 + 2} + 6\left(\frac{-1}{2}\right) \stackrel{?}{=} \left(\frac{-1}{2}\right) - 4$$

$$\frac{3}{2} - 3 \stackrel{?}{=} \frac{-9}{2}$$

$$\frac{-3}{2} \neq \frac{-9}{2}$$

Hence, $-\frac{1}{2}$ is not a root of the equation, but $\frac{-7}{6}$ is a root.

Problem Solving Examples:

Solve for x: $4x^2 - 7 = 0$.

This quadratic equation can be solved for x using the quadratic formula, which applies to equations in the form $ax^2 + bx + c = 0$ (in our equation $b = 0$). There is, however, an easier method that we can use:

Adding 7 to both sides, $4x^2 = 7$

Dividing both sides by 4, $x^2 = \frac{7}{4}$

Taking the square root of both sides, $x = \pm\sqrt{\frac{7}{4}} = \pm\frac{\sqrt{7}}{2}$.

The double sign ± (read "plus or minus") indicates that the two roots of the equation are $+\frac{\sqrt{7}}{2}$ and $-\frac{\sqrt{7}}{2}$.

Solve the equation $2x^2 - 5x + 3 = 0$.

A $2x^2 - 5x + 3 = 0$

The equation is a quadratic equation of the form $ax^2 + bx + c = 0$ in which $a = 2$, $b = -5$, and $c = 3$. Therefore, the quadratic formula $x = \frac{-b \pm \sqrt{b^2 - 4ac}}{2a}$ may be used to find the solutions of the given equation. Substituting the values for a, b, and c in the quadratic formula:

$$x = \frac{-(-5) \pm \sqrt{(-5)^2 - 4(2)(3)}}{2(2)}$$

$$x = \frac{5 \pm \sqrt{1}}{4}$$

$$x = \frac{5+1}{4} = \frac{3}{2} \text{ and } x = \frac{5-1}{4} = 1$$

Check: Substituting $x = \frac{3}{2}$ in the given equation,

$$2\left(\frac{3}{2}\right)^2 - 5\left(\frac{3}{2}\right) + 3 = 0$$
$$0 = 0$$

Substituting $x = 1$ in the given equation,

$$2(1)^2 - 5(1) + 3 = 0$$
$$0 = 0$$

So the roots of $2x^2 - 5x + 3 = 0$ are $x = \frac{3}{2}$ and $x = 1$.

9.3 Quadratic Functions

The function $f(x) = ax^2 + bx + c$, $a \pm 0$, where a, b, and c are real numbers, is called a quadratic function (or a function of second degree) in one unknown.

The graph of $y = ax^2 + bx + c$ is a curve known as a parabola.

The vertex of the parabola is the point $v\left(\frac{-b}{2a}, \frac{4ac - b^2}{4a}\right)$. The parabola's axis is the line $x = \frac{-b}{2a}$.

The graph of the parabola opens upward if $a > 0$ and downward if $a < 0$. If $a = 0$ the quadratic is reduced to a linear function whose graph is a straight line.

Figures 9.1 and 9.2 show parabolas with $a > 0$, and $a < 0$, respectively.

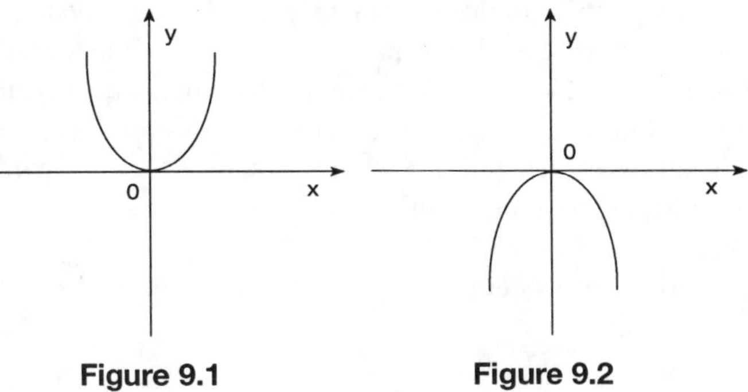

Figure 9.1

Figure 9.2

Problem Solving Examples:

Solve the system

$$y = -x^2 + 7x - 5 \qquad (1)$$
$$y - 2x = 2 \qquad (2)$$

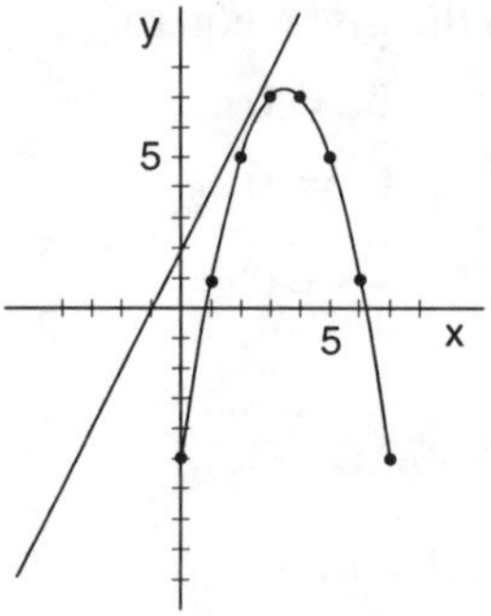

Figure 9.3

Solving Equation (2) for y yields an expression for y in terms of x. Substituting this expression in Equation (1),

$$2x + 2 = -x^2 + 7x - 5 \qquad (3)$$

We have a single equation, in terms of a single variable, to be solved. Writing Equation (3) in standard quadratic form,

$$x^2 - 5x + 7 = 0 \qquad (4)$$

Since the equation is not factorable, the roots are not found in this manner. Evaluating the discriminant will indicate whether Equation (4) has real roots. The discriminant, $b^2 - 4ac$, of Equation (4) equals $(-5)^2 - 4(1)(7) = 25 - 28 = -3$. Since the discriminant is negative, equation (4) has no real roots, and therefore the system has no real solution. In terms of the graph, Figure 9.3 shows that the parabola and the straight line have no point in common.

Solve the system

$$y = 3x^2 - 2x + 5 \qquad (1)$$
$$y = 4x + 2 \qquad (2)$$

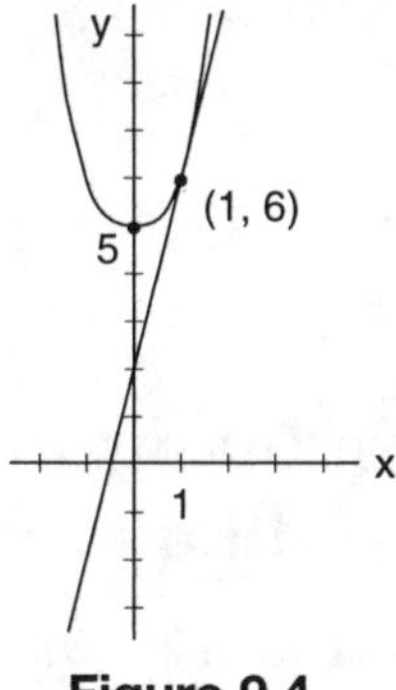

Figure 9.4

To obtain a single equation with one unknown variable, x, substitute the value of y from Equation (2) in Equation (1),

$$4x + 2 = 3x^2 - 2x + 5. \qquad (3)$$

Writing Equation (3) in standard quadratic form,

$$3x^2 - 6x + 3 = 0. \qquad (4)$$

We may simplify equation (4) by dividing both members by 3, which is a factor common to each term:

$$x^2 - 2x + 1 = 0. \qquad (5)$$

To find the roots, factor and set each factor = 0. This may be done since a product = 0 implies one or all of the factors must = 0.

$$(x - 1)(x - 1) = 0 \qquad (6)$$

$$\begin{array}{r|r} x - 1 = 0 & x - 1 = 0 \\ x = 1 & x = 1 \end{array}$$

Equation (5) has two equal roots, each equal to 1. For $x = 1$, from Equation (2), we have $y = 4(1) + 2 = 6$. Therefore, the system has but one common solution:

$$x = 1, \quad y = 6.$$

Figure 9.4 indicates that our solution is probably correct. We may also check to see if our values satisfy Equation (1) as well:

Substituting in:

$$\begin{aligned} y &= 3x^2 - 2x + 5 \\ 6 &\stackrel{?}{=} 3(1)^2 - 2(1) + 5 \\ 6 &\stackrel{?}{=} 3 - 2 + 5 \\ 6 &= 6 \end{aligned}$$

9.4 Quadratic Equations in Two Unknowns and Systems of Equations

A quadratic equation in two unknowns has the general form:

$$ax^2 + bxy + cy^2 + dx + ey + f = 0$$

where a, b, and c are not all zero and a, b, c, d, e, and f are constants.

Graphing: If $b^2 - 4ac < 0$, $b \neq 0$ and $a \neq c$, the graph of $ax^2 + bxy + cy^2 + dx + ey + f$ is a closed curve called an ellipse. If $b = 0$ and $a = c$, the graph $ax^2 + bxy + cy^2 + dx + ey + f$ is a point or a circle, or else it does not exist.

If $b^2 - 4ac > 0$, the graph of $ax^2 + bxy + cy^2 + dx + ey + f = 0$ is a curve called hyperbola or two intersecting lines.

If $b^2 - 4ac = 0$, the graph of $ax^2 + bxy + cy^2 + dx + ey + f = 0$ is a parabola or a pair of parallel lines which may be coincident, or else it does not exist.

9.5 Solving Systems of Equations Involving Quadratics

Some methods for solving systems of equations involving quadratics are given below:

A) One linear and one quadratic equation

Solve the linear equation for one of the two unknowns, then substitute this value into the quadratic equation.

B) Two quadratic equations

Eliminate one of the unknowns using the method given for solving systems of linear equations.

Example:

$$\begin{cases} x^2 + y^2 = 9 & (1) \\ x^2 + 2y^2 = 18 & (2) \end{cases}$$

Subtracting Equation (1) from (2), we obtain:

$$y^2 = 9, \ \ y = \pm 3$$

By substituting the values of y into (1) or (2), we obtain:

$$x_1 = 0 \text{ and } x_2 = 0$$

So the solutions are:

$$x = 0, y = 3 \text{ and } x = 0, y = -3$$

C) Two quadratic equations, one homogeneous

An equation is said to be homogeneous if it is of the form

$$ax^2 + bxy + cy^2 + dx + ey = 0.$$

Consider the system

$$\begin{cases} x^2 + 3xy^2 + 2y^2 = 0 & (1) \\ x^2 - 3xy + 2y^2 = 12 & (2) \end{cases}$$

Equation (1) can be factored into the product of two linear equations:

$$x^2 + 3xy + 2y^2 = (x + 2y)(x + y) = 0$$

From this we determine that:

$$\begin{aligned} x + 2y &= 0 \Rightarrow x = -2y \\ x + \ \ y &= 0 \Rightarrow x = -y \end{aligned}$$

Substituting $x = -2y$ into Equation (2), we find:

$$\begin{aligned} (-2y)^2 - 3(-2y)y + 2y^2 &= 12 \\ 4y^2 + 6y^2 + 2y^2 &= 12 \\ 12y^2 &= 12 \end{aligned}$$

$$y^2 = 1$$
$$y = \pm 1, \text{ so } x = \pm 2$$

Substituting $x = -y$ into Equation (2) yields:

$$(-y)^2 - 3(-y)y + 2y^2 = 12$$
$$y^2 + 3y^2 + 2y^2 = 12$$
$$6y^2 = 12$$
$$y^2 = 2$$
$$y = \pm\sqrt{2}, \text{ so } x = \pm\sqrt{2}$$

So the solutions of Equations (1) and (2) are:

$$x = 2, y = -1, x = -2, y = 1, x = \sqrt{2}, y = -\sqrt{2},$$
$$\text{and } x = -\sqrt{2}, y = \sqrt{2}$$

D) Two quadratic equations of the form

$$ax^2 + bxy + cy^2 = d$$

Combine the two equations to obtain a homogeneous quadratic equation then solve the equations by the third method.

E) Two quadratic equations, each symmetrical in x and y

Note: An equation is said to be symmetrical in x and y if by exchanging the coefficients of x and y we obtain the same equation.

Example: $x^2 + y^2 = 9$.

To solve systems involving this type of equations, substitute $u + v$ for x and $u - v$ for y and solve the resulting equations for u and v.

Example: Given the system below:

$$\begin{cases} x^2 + y^2 = 25 & (1) \\ x^2 + xy + y^2 = 37 & (2) \end{cases}$$

Substitute:

$$x = u + v$$
$$y = u - v$$

If we substitute the new values for x and y into Equation (2) we obtain:

$$(u + v)^2 + (u + v)(u - v) + (u - v)^2 = 37$$
$$u^2 + 2uv + v^2 + u^2 - v^2 + u^2 - 2uv + v^2 = 37$$
$$3u^2 + v^2 = 37.$$

If we substitute for x and y into Equation (1), we obtain:

$$(u + v)^2 + (u - v)^2 = 25$$
$$u^2 + 2uv + v^2 + u^2 - 2uv + v^2 = 25$$
$$2u^2 + 2v^2 = 25.$$

The "new" system is:

$$\begin{cases} 3u^2 + v^2 = 37 \\ 2u^2 + 2v^2 = 25 \end{cases}$$

By substituting $a = u^2$ and $b = v^2$, these equations become:

$$\begin{cases} 3a + b = 37 \\ 2a + 2b = 25 \end{cases}$$

and

$$a = \frac{49}{4}, \qquad b = \frac{1}{4}.$$

So

$$u^2 = \frac{49}{4} \qquad \text{and} \qquad v^2 = \frac{1}{4}$$

$$u = \pm\frac{7}{2}$$

$$v = \pm\frac{1}{2}$$

$$x = \frac{7}{2} + \frac{1}{2} = 4 \qquad \text{or} \qquad \frac{-7}{2} - \frac{1}{2} = -4$$

$$y = \frac{7}{2} - \frac{1}{2} = 3 \qquad \text{or} \qquad \frac{-7}{2} + \frac{1}{2} = -3$$

Since x and y are symmetrical, the possible solutions are (4, 3), (– 4, –3), (3, 4), (–3, – 4).

Note that if the equation is symmetrical it is possible to interchange the solutions too. If $x = 3$, then $y = 4$ or vice-versa.

Problem Solving Examples:

Solve the system

$$\begin{aligned} 2x^2 - 3xy - 4y^2 + x + y - 1 &= 0 \\ 2x - y &= 3. \end{aligned}$$

A system of equations consisting of one linear and one quadratic is solved by expressing one of the unknowns in the linear equation in terms of the other, and substituting the result in the quadratic equation. From the second equation, $y = 2x - 3$. Replacing y by this linear function of x in the first equation, we find

$$2x^2 - 3x(2x - 3) - 4(2x - 3)^2 + x + 2x - 3 - 1 = 0$$

$$2x^2 - 3x(2x - 3) - 4(4x^2 - 12x + 9) + x + 2x - 3 - 1 = 0$$

Distribute, $\quad 2x^2 - 6x^2 + 9x - 16x^2 + 48x - 36 + x + 2x - 3 - 1 = 0$

Combine terms, $\quad -20x^2 + 60x - 40 = 0$

Divide both sides by –20, $\quad \dfrac{-20x^2}{-20} + \dfrac{60x}{-20} - \dfrac{40}{-20} = \dfrac{0}{-20}$

$$x^2 - 3x + 2 = 0$$

Factoring, $\quad (x - 2)(x - 1) = 0$

Setting each factor equal to zero, we obtain:

$$\begin{aligned} x - 2 &= 0 \\ x &= 2 \end{aligned} \qquad\qquad \begin{aligned} x - 1 &= 0 \\ x &= 1 \end{aligned}$$

To find the corresponding y-values, substitute the x-values in $y = 2x - 3$:

$$\begin{aligned} \text{when } x &= 1, & \text{when } x &= 2, \\ y &= 2(1) - 3 & y &= 2(2) - 3 \\ y &= 2 - 3 & y &= 4 - 3 \\ y &= -1 & y &= 1 \end{aligned}$$

Therefore, the two solutions of the system are

$$(1, -1), \qquad (2, 1),$$

and the solution set is $\{(1, -1), (2, 1)\}$.

Solve the system

$$2x^2 - 3xy + 4y^2 = 3 \qquad (1)$$
$$x^2 + xy - 8y^2 = -6 \qquad (2)$$

Multiply both sides of the first equation by 2.

$$2(2x^2 - 3xy + 4y^2) = 2(3)$$
$$4x^2 - 6xy + 8y^2 = 6 \qquad (3)$$

Add Equation (3) to Equation (2):

$$x^2 + xy - 8y^2 = -6$$
$$4x^2 - 6xy + 8y^2 = 6$$

$$5x^2 - 5xy = 0 \qquad (4)$$

Factoring out the common factor, $5x$, from the left side of Equation (4):

$$5x(x - y) = 0$$

Whenever a product $ab = 0$, where a and b are any two numbers, either $a = 0$ or $b = 0$ or both. Hence, either

$$\begin{aligned} 5x &= 0 & \text{or} \qquad x - y &= 0 \\ x &= 0/5 & x &= y \\ x &= 0 \end{aligned}$$

Substituting $x = 0$ in Equation (1):

$$2(0)^2 - 3(0)y + 4y^2 = 3$$
$$0 - 0 + 4y^2 = 3$$
$$4y^2 = 3$$
$$y^2 = \frac{3}{4}$$

$$y = \pm\sqrt{\frac{3}{4}}$$
$$= \pm\frac{\sqrt{3}}{\sqrt{4}}$$
$$= \pm\frac{\sqrt{3}}{2}$$

Hence, two solutions are: $\left(0, \frac{\sqrt{3}}{2}\right), \left(0, -\frac{\sqrt{3}}{2}\right)$.

Substituting x for y $(x = y)$ in equation (1):

$$2x^2 - 3x(x) + 4(x)^2 = 3$$
$$2x^2 - 3x^2 + 4x^2 = 3$$
$$-x^2 + 4x^2 = 3$$
$$3x^2 = 3$$
$$x^2 = 3/3$$
$$x^2 = 1$$
$$x = \pm\sqrt{1} = \pm 1$$

Therefore, when $x = 1$, $y = x = 1$. Also, when $x = -1$, $y = x = -1$. Hence, two other solutions are: $(1, 1)$ and $(-1, -1)$. Thus the four solutions of the system are

$$\left(0, \frac{\sqrt{3}}{2}\right), \left(0, -\frac{\sqrt{3}}{2}\right), (1, 1), \text{ and } (-1, -1).$$

Solve the system

$$3x^2 + 4y^2 = 8 \qquad (1)$$

$$x^2 - y^2 = 5 \qquad (2)$$

Substituting u for x^2 and v for y^2 leads to the system of linear equations

$$3u + 4v = 8 \qquad (3)$$

$$u - v = 5 \qquad (4)$$

Multiplying both sides of Equation (4) by 3, we obtain:

$$3(u - v) = 3(5)$$

$$3u - 3v = 15 \qquad (5)$$

Subtracting Equation (5) from Equation (3):

$$\begin{aligned} 3u + 4v &= 8 \\ -(3u - 3v &= 15) \\ \hline 7v &= -7 \end{aligned}$$

Dividing both sides by 7:

$$\frac{7v}{7} = \frac{-7}{7}$$

$$v = -1$$

Since $v = y^2$,

$$v = -1 = y^2$$

$$\pm\sqrt{-1} = y$$

Since $i^2 = -1$ or $i = \sqrt{-1}$

$$\pm i = y.$$

Substituting the value $y = i$ into Equation (2):

$$x^2 - (i)^2 = 5$$

$$x^2 - i^2 = 5$$

$$x^2 - (-1) = 5$$
$$x^2 + 1 = 5$$

Subtracting 1 from both sides:

$$x^2 + \cancel{1} - \cancel{1} = 5 - 1$$
$$x^2 = 4$$
$$x = \pm\sqrt{4} = \pm 2$$

Hence, two solutions of the original system of equations are:

$$(2, i), (-2, i)$$

Substituting the value $y = -i$ into Equation (2):

$$x^2 - (-i)^2 = 5$$
$$x^2 - (i^2) = 5$$
$$x^2 - (-1) = 5$$
$$x^2 + 1 = 5$$

Subtracting 1 from both sides:

$$x^2 + \cancel{1} - \cancel{1} = 5 - 1$$
$$x^2 = 4$$
$$x = \pm\sqrt{4} = \pm 2$$

Hence, two other solutions of the original system of equations are:

$$(2, -i), (-2, -i).$$

Therefore, the solution set of the original system of equations is:

$$\{(2, i), (-2, i), (2, -i), (-2, -i)\}.$$

Other systems that involve quadratic equations may be solved by replacing the given system with an equivalent system that is easier to solve.

Quiz: Quadratic Equations

Solve for all values of x.

1. $x^2 - 2x - 8 = 0$

   (A) 4 and – 2
   (B) 4 and 8
   (C) 4
   (D) – 2 and 8
   (E) – 2

2. $x^2 + 2x - 3 = 0$

   (A) – 3 and 2
   (B) 2 and 1
   (C) 3 and 1
   (D) – 3 and 1
   (E) – 3

3. $x^2 - 7x = -10$

   (A) – 3 and 5
   (B) 2 and 5
   (C) 2
   (D) – 2 and – 5
   (E) 5

4. $x^2 - 8x + 16 = 0$

   (A) 8 and 2
   (B) 1 and 16
   (C) 4
   (D) – 2 and 4
   (E) 4 and – 4

5. $3x^2 + 3x = 6$

   (A) 3 and – 6
   (B) 2 and 3
   (C) – 3 and 2
   (D) 1 and – 3
   (E) 1 and – 2

6. $x^2 + 7x = 0$

(A) 7
(B) 0 and – 7
(C) – 7
(D) 0 and 7
(E) 0

7. $x^2 - 25 = 0$

(A) 5
(B) 5 and – 5
(C) 15 and 10
(D) – 5 and 10
(E) – 5

8. $2x^2 + 4x = 16$

(A) 2 and – 2
(B) 8 and – 2
(C) 4 and 8
(D) 2 and – 4
(E) 2 and 4

9. $2x^2 - 11x - 6 = 0$

(A) 1 and – 3
(B) 0 and 4
(C) 1
(D) – 4
(E) $-^1/_2$ and 6

10. $x^2 - 2x - 3 = 0$

(A) 0
(B) –1 and 3
(C) 5 and –3
(D) 2
(E) 1 and – 2

ANSWER KEY

1.	(A)	6.	(B)
2.	(D)	7.	(B)
3.	(B)	8.	(D)
4.	(C)	9.	(E)
5.	(E)	10.	(B)

CHAPTER 10

Equations of Higher Order

10.1 Methods to Solve Equations of Higher Order

A) Factorization:

Given: $x^4 - x = 0$

By factorization, it is possible to express this equation as:

$x(x^3 - 1) = 0$

The equation above can still be factored to give:

$x(x - 1)(x^2 + x + 1) = 0,$

which means that x, $(x - 1)$, or $(x^2 + x + 1)$ must be equal to zero.

$x = 0$ means 0 is a root of $x^4 - x = 0$.

$x - 1 = 0$ means 1 is a root.

To solve $x^2 + x + 1 = 0$, we can use the quadratic formula:

$$\frac{-1 \pm \sqrt{1^2 - 4(1)(1)}}{2(1)} = \frac{-1 \pm \sqrt{-3}}{2} = \frac{-1 \pm \sqrt{3}i}{2}$$

This implies $\frac{-1+\sqrt{3}i}{2}$ and $\frac{-1-\sqrt{3}i}{2}$ are roots.

So $\frac{-1+\sqrt{3}i}{2}$ and $\frac{-1-\sqrt{3}i}{2}$ are solutions of $x^2 + x + 1 = 0$ and therefore of $x^4 - x = 0$. This means the solution set of $x^4 - x = 0$ is:

$$\left\{0,\ 1,\ \frac{-1+\sqrt{3i}}{2},\ \frac{-1-\sqrt{3i}}{2}\right\}$$

B) If the equation to be solved is of third degree, it is possible to write it as:

$$x^3 + b_1x^2 + b_2x + b_3 = 0$$

where $-b_1$ = sum of the roots,
b_2 = sum of the products of the roots taken two at a time,
and
$(-1)^3b_3$ = product of the roots.

C) It is possible to determine the roots of an equation by writing the equation in the form: $y = f(x)$ and checking the values of x for which y is zero. These values of x are called zeros of the function and correspond to the roots. Figure 10.1 shows this procedure:

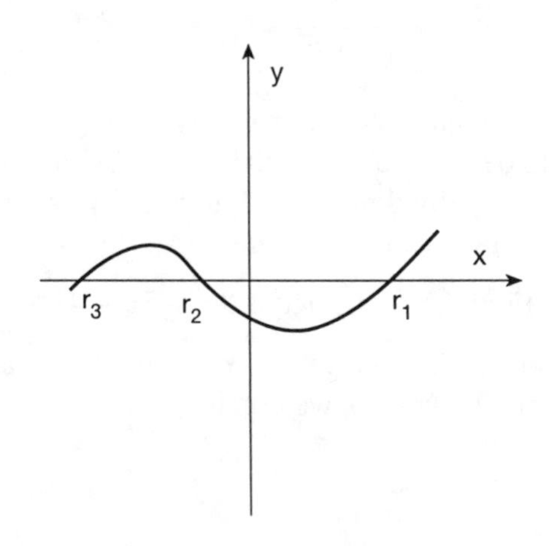

Figure 10.1

r_1, r_2, and r_3 can be determined by graphing the function $f(x)$ at a large number of points and connecting them. Note that the values read are just approximations of the roots.

D) If the equation to be solved is of fourth degree, we can replace x^2 with y and solve a second degree equation using the quadratic formula. The square root of y would give the value of x. Note: This is only possible if the coefficients of the x^3-term and the x-term are zero.

Example: Solve $x^4 - 12x^2 + 8 = 0$

$$x^4 - 12x^2 + 8 = (x^2)^2 - 12x^2 + 8 = 0$$

Let $x^2 = y$:

$$y^2 - 12y + 8 = 0$$

$$y = \frac{12 \pm \sqrt{144 - 4(1)(8)}}{2(1)}$$

$$y = 6 \pm 2\sqrt{7}$$

$$x = \pm\sqrt{6 \pm 2\sqrt{7}}$$

So the four roots are:

$$+\sqrt{6+2\sqrt{7}},\ -\sqrt{6+2\sqrt{7}},\ +\sqrt{6-2\sqrt{7}},\ -\sqrt{6-2\sqrt{7}}$$

E) Inspection: Given an equation of any order greater than two; if it is possible to determine one of the roots r_1 by inspection, then $(x - r_1)$ is a factor. By dividing the polynomial by $(x - r_1)$, we can find other roots by factoring the quotient.

Example: Find the roots of $x^4 + 2x^3 - 5x^2 - 4x + 6 = 0$.

By inspection, $x = 1$ is a root. So $(x - 1)$ is a factor and by dividing the polynomial by $(x - 1)$, we obtain:

$$
\begin{array}{r l}
 & x^3 + 3x^2 - 2x - 6 \\
x-1 \,\big) & x^4 + 2x^3 - 5x^2 - 4x + 6 \\
 & -(x^4 - x^3) \\
\hline
 & 3x^3 - 5x^2 \\
 & -(3x^3 - 3x^2) \\
\hline
 & -2x^2 - 4x \\
 & -(-2x^2 + 2x) \\
\hline
 & (-6x + 6) \\
 & -(-6x + 6) \\
\hline
 & 0
\end{array}
$$

So we get $(x - 1)(x + 3)(x^2 - 2) = 0$. The roots of $x^4 + 2x^3 - 5x^2$ $4x + 6 = 0$ are $x = 1$, $x = -3$, and $x = \pm\sqrt{2}$.

Problem Solving Examples:

Find all solutions of the equation $x^3 - 3x^2 - 10x = 0$.

Factor out the common factor of x from the terms on the left side of the given equation. Therefore,

$$x^3 - 3x^2 - 10x = x(x^2 - 3x - 10) = 0.$$

Whenever $ab = 0$ where a and b are any two numbers, either $a = 0$ or $b = 0$. Hence, either $x = 0$ or $x^2 - 3x - 10 = 0$. The expression $x^2 - 3x - 10$ factors into $(x - 5)(x + 2)$. Therefore, $(x - 5)(x + 2) = 0$. Applying the above law again:

$$
\begin{array}{llll}
\text{either} & x - 5 = 0 & \text{or} & x + 2 = 0 \\
 & x = 5 & & x = -2
\end{array}
$$

Hence,

$$x^3 - 3x^2 - 10x = x(x - 5)(x + 2) = 0$$

Either $x = 0$ or $x = 5$ or $x = -2$. The solution set is $X = \{0, 5, -2\}$.

We have shown that, if there is a number x such that $x^3 - 3x^2 - 10x = 0$, then $x = 0$ or $x = 5$ or $x = -2$. Finally, to see that these three numbers are actually solutions, we substitute each of them in turn in the original equation to see whether or not it satisfies the equation $x^3 - 3x^2 - 10x = 0$.

Check: Replacing x by 0 in the original equation,

$$(0)^3 - 3(0)^2 - 10(0) = 0 - 0 - 0 = 0.$$

Replacing x by 5 in the original equation,

$$\begin{aligned}(5)^3 - 3(5)^2 - 10(5) &= 125 - 3(25) - 50\\ &= 125 - 75 - 50\\ &= 50 - 50 = 0.\end{aligned}$$

Replacing x by -2 in the original equation,

$$\begin{aligned}(-2)^3 - 3(-2)^2 - 10(-2) &= -8 - 3(4) + 20\\ &= -8 - 12 + 20\\ &= -20 + 20 = 0.\end{aligned}$$

Q Graph the function $y = x^3 - 9x$.

Choosing values of x in the interval $-4 \le x \le 4$, we have for $y = x^3 - 9x$,

x	-4	-3	-2	-1	0	1	2	3	4
y	-28	0	10	8	0	-8	-10	0	28

Notice that for each ordered pair (x, y) listed in the table there exists a pair $(-x, -y)$ which also satisfies the equation, indicating symmetry with respect to the origin. To prove that this is true for all points on the curve, we substitute $(-x, -y)$ for (x, y) in the given equation and show that the equation is unchanged. Thus,

$$-y = (-x)^3 - 9(-x) = -x^3 + 9x$$

or, multiplying each member by –1,

$$y = x^3 - 9x$$

which is the original equation.

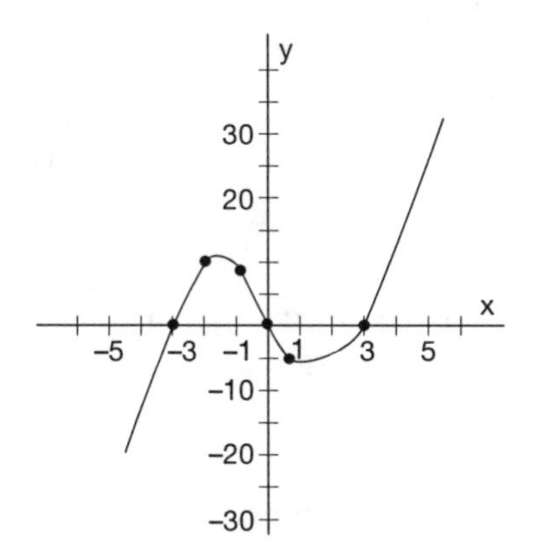

Figure 10.2

The curve is illustrated in Figure 10.2. The domain and range of the function have no restrictions in the set of real numbers. The x-intercepts are found from

$$\begin{aligned} y = 0 &= x^3 - 9x \\ 0 &= x(x^2 - 9) \\ 0 &= x(x - 3)(x + 3) \end{aligned}$$

$$x = 0 \quad\Big|\quad \begin{aligned} x - 3 &= 0 \\ x &= 3 \end{aligned} \quad\Big|\quad \begin{aligned} x + 3 &= 0 \\ x &= -3 \end{aligned}$$

The curve has three x-intercepts at $x = -3$, $x = 0$, and $x = 3$. This agrees with the fact that a cubic equation has three roots. The curve has a single y-intercept at $y = 0$ since for $x = 0$, $y = 0^3 - 9(0) = 0$.

10.2 Theory of Equations

A) Remainder Theorem: If a is any constant and if the polynomial $P(x)$ is divided by $(x - a)$, the remainder is $P(a)$.

Example: Given a polynomial $2x^3 - x^2 + x + 4$ divided by $x - 1$, the remainder is $2(1)^3 - (1)^2 + 1 + 4 = 6$. That is

$$2x^2 - x^2 + x + 4 = q(x) + \frac{6}{(x-1)}$$

where $q(x)$ is a polynomial.

Note that in this case $a = 1$.

B) Factor Theorem: If a is a root of the equation $f(x) = 0$, then $(x - a)$ is a factor of $f(x)$.

C) Synthetic Division: This method allows us to check if a certain constant c is a root of the given polynomial and, if it is not, it gives us the remainder of the division by $(x - c)$.

The general polynomial

$$P(x) = a_n x^n + a_{n-1} x^{n-1} + \ldots + a_i x^i + \ldots + a_0 x^0$$

can be represented by its coefficients a_i, written in descending powers of x.

The method consists of the following steps:

(a) Write the coefficients a_i of the polynomial.

(b) Multiply the first coefficient by the additive inverse of the constant term in the divisor and add it to the following coefficient of $p(x)$.

(c) Continue until the last coefficient of $p(x)$ is reached. The resulting numbers are the coefficients of the quotient polynomial, with the last number representing the remainder. If the remainder is 0, then c is a root of $p(x)$.

Given $x^4 + 6x^3 - 2x^2 + 5$, divide by $x - (-1)$.

$$\frac{x^4 + 6x^3 - 2x^2 + 5}{x + 1} \Rightarrow$$

$\boxed{1}$	6	–2	0	5	–1
	–1				
1	$\boxed{5}$	–2	0	5	–1
		–5			
1	5	$\boxed{-7}$	0	5	–1

Note that this can be written as: $x^4 + 6x^3 - 2x^2 + 0x + 5$, which explains the zero in the synthetic division.

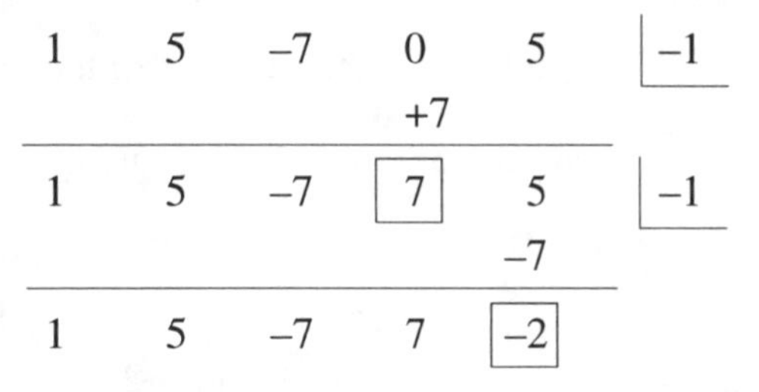

So the remainder is –2 and:

$$\frac{x^4 + 6x^3 - 2x^2 + 5}{x+1} = x^3 + 5x^2 - 7x + 7 - \frac{2}{x+1}$$

Given $x^3 - 7x - 6$, check if 3 is a root:

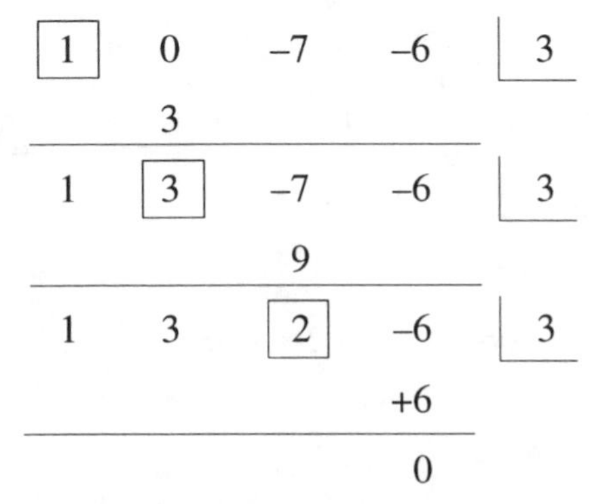

Yes, since the remainder is 0. Note that here x^2 has coefficient zero, which explains the zero in the synthetic division.

Problem Solving Examples:

Form the equation of the fourth degree with rational coefficients, one of whose roots is $\sqrt{2} + \sqrt{-3}$.

Here we must have $\sqrt{2} + \sqrt{-3}$, $\sqrt{2} - \sqrt{-3}$ as one pair of roots, and $-\sqrt{2} + \sqrt{-3}$, $-\sqrt{2} - \sqrt{-3}$ as another pair.

Recall that every quadratic equation may be written in the form:

$ax^2 + bx + c = 0$

Also, recall the well-known formula that the sum of the roots of a quadratic equation is $\frac{-b}{a}$, and the product of the roots is $\frac{c}{a}$. Thus, finding the sum and product of our known roots will give us their corresponding equation.

Our first pair of roots are $\sqrt{2}+\sqrt{-3}$ and $\sqrt{2}-\sqrt{-3}$. Their sum is:

$\sqrt{2}+\sqrt{-3}+(\sqrt{2}-\sqrt{-3}) = 2\sqrt{2}$, and their product is:

$$\begin{aligned}(\sqrt{2}+\sqrt{-3})(\sqrt{2}-\sqrt{-3}) &= 2+\sqrt{2}\sqrt{-3}-\sqrt{2}\sqrt{-3}-\sqrt{-3}\sqrt{-3}\\ &= 2-(-3)\\ &= 5\end{aligned}$$

So our sum is $2\sqrt{2} = \frac{-b}{a}$, and our product is $5 = \frac{c}{a}$. Therefore, $a = 1$, $b = -2\sqrt{2}$, $c = 5$, and the quadratic factor corresponding to this pair of roots is $x^2 - 2\sqrt{2}\,x + 5$.

The second pair of roots are $-\sqrt{2}+\sqrt{-3}$ and $-\sqrt{2}-\sqrt{-3}$. Their sum is:

$-\sqrt{2}+\sqrt{-3}+(-\sqrt{2}-\sqrt{-3}) = -2\sqrt{2}$

and their product is:

$$\begin{aligned}(-\sqrt{2}+\sqrt{-3})(-\sqrt{2}-\sqrt{-3}) &= 2-\sqrt{2}\sqrt{-3}+\sqrt{2}\sqrt{-3}-\sqrt{-3}\sqrt{-3}\\ &= 2-(-3)\\ &= 2+3\\ &= 5.\end{aligned}$$

So our sum is $-2\sqrt{2} = \frac{-b}{a}$, and our product is $5 = \frac{c}{a}$. Therefore, $a = 1$, $b = 2\sqrt{2}$, $c = 5$, and the quadratic factor corresponding to the second pair of roots is $x^2 + 2\sqrt{2}\,x + 5$.

Thus, the required equation is

$$(x^2 + 2\sqrt{2}x + 5)(x^2 - 2\sqrt{2}x + 5) = 0,$$

or $$x^4 + 2x^2 + 25 = 0.$$

Solve the equation $6x^4 - 13x^3 - 35x^2 - x + 3 = 0$, having given that one root is $2 - \sqrt{3}$.

Since $2 - \sqrt{3}$ is a root, $2 + \sqrt{3}$ is also a root. Recall that every quadratic equation may be written in the form:

$$ax^2 + bx + c = 0.$$

Also, recall the well-known formula that the sum of the roots of a quadratic equation is $\frac{-b}{a}$, and the product of the roots is $\frac{c}{a}$. Thus, finding the sum and product of our known roots will give us their corresponding equation.

Our roots are $2 - \sqrt{3}$ and $2 + \sqrt{3}$. Their sum is $2 - \sqrt{3} + (2 + \sqrt{3}) = 4$ and their product is:

$$(2 - \sqrt{3})(2 + \sqrt{3}) = 4 - 2\sqrt{3} + 2\sqrt{3} - 3 = 1.$$

So our sum is $4 = \frac{-b}{a}$, and our product is $1 = \frac{c}{a}$. Therefore, $a = 1$, $b = -4$, $c = 1$, and the equation is $x^2 - 4x + 1$.

Now, dividing this new equation into the given equation, we obtain:

$$\begin{array}{r|l} & 6x^2 + 11x + 3 \\ \hline x^2 - 4x + 1 & 6x^4 - 13x^3 - 35x^2 - x + 3 \\ & 6x^4 - 24x^3 + 6x^2 \\ \hline & \quad 11x^3 - 41x^2 - x + 3 \\ & \quad 11x^3 - 44x^2 + 11x \\ \hline & \qquad 3x^2 - 12x + 3 \\ & \qquad 3x^2 - 12x + 3 \\ \hline & \qquad\qquad 0 \end{array}$$

Thus, $6x^4 - 13x^3 - 35x^2 - x + 3 = (x^2 - 4x + 1)(6x^2 + 11x + 3)$.

Since $6x^4 - 13x^3 - 35^2 - x + 3 = 0$, then $(x^2 - 4x + 1)(6x^2 + 11x + 3) = 0$, and this means that $6x^2 + 11x + 3 = 0$.

Factoring this equation, we obtain $(3x + 1)(2x + 3) = 0$. Thus, $3x + 1 = 0$ or $x = -\frac{1}{3}$; and $2x + 3 = 0$, or $x = -\frac{3}{2}$. Therefore, the roots of the given equation are $-\frac{1}{3}$, $-\frac{3}{2}$, $2 + \sqrt{3}$, and $2 - \sqrt{3}$.

10.3 Algebraic Theorems

A) Every polynomial equation $f(x) = 0$ of degree greater than zero has at least one root either real or complex. This is known as the fundamental theorem of algebra.

B) Every polynomial equation of degree n has exactly n roots.

C) If a polynomial equation $f(x) = 0$ with real coefficients has a root $a + bi$, then the conjugate of this complex number $a - bi$ is also a root of $f(x) = 0$.

D) If $a + \sqrt{b}$ is a root of the polynomial equation $f(x) = 0$ with rational coefficients, then $a - \sqrt{b}$ is also a root, where a and b are rational and $\sqrt{b}$ is irrational.

E) If a rational fraction in lowest terms is a root of the equation

$$a_n x^n + a_{n-1} x^{n-1} + \ldots + a_1 x + a_0 = 0,$$

$a_0 \neq 0$, and the a_i are integers, then b is a factor of a and c is a factor of a_n.

Furthermore, any rational roots of the equation below must be integers and factors of q_n.

$$x^n + q_1 x^{n-1} + q_2 x^{n-2} + \ldots + q_{n-1} x + q_n = 0$$

Note that q_1, q_2, …, q_n are integers.

Given

$$f(x) = a_n x^n + a_{n-1} x^{n-1} + \ldots + a_0 = 0$$

where a_n, a_{n-1}, ..., a_0 are real and $a_n > 0$, then q is an upper limit for all real roots of $f(x) = 0$ (a number q is called an upper limit for the real roots of $f(x) = 0$ if none of the roots is greater than q) if upon synthetic division of $f(x)$ by $x - q$, all of the numbers obtained in the last row* have the same sign. If, however, upon synthetic division of $f(x)$ by $x - p$, all of the numbers obtained in the last row* have alternating signs, then p is a lower limit for all the real roots of $f(x) = 0$. A number p is called a lower limit for the real roots if none of the roots is less than p.

*Note that last row refers to the final line obtained by a synthetic division and corresponds to the line that gives the remainder.

F) Given a general polynomial of the form below:

$$f(x) = x^n + p_1x^{n-1} + p_2x^{n-2} + \dots + p_{n-1}x + p_n = 0$$

it has the following properties:

a) $-p_1$ = sum of the roots.

b) p_2 = sum of the products of the roots taken two at a time.

c) $-p_3$ = sum of the products of the roots taken three at a time.

d) $(-1)^n p_n$ = product of all the roots of $f(x) = 0$.

10.4 Descartes' Rule of Signs

Variation in sign: A polynomial $f(x)$ with real coefficients is said to have a variation in sign if, after arranging its terms in descending powers of x, two successive terms differ in sign.

Example: $3x^5 - 4x^4 + 3x^3 - 9x^2 - x + 1$ has four variations.

10.4.1 Descartes' Rule of Signs

The number of positive roots of a polynomial equation $f(x) = 0$ with real coefficients cannot exceed the number of variations in sign of $f(x)$. The difference between the number of variations and the number of positive roots of the equation is an even number.

The number of negative roots of $f(x) = 0$ cannot exceed the number of variations of sign of $f(-x)$. The difference between the number of variations and the number of negative roots is an even number.

Example: $3x^5 - 4x^4 + 3x^3 - x + 1 = 0$ has four variations in sign so the number of positive roots cannot exceed four. It can be 0, 2, or 4. $f(-x)$ would be obtained as shown below:

$$3(-x)^5 - 4(x)^4 + 3(-x)^3 - (-x) + 1 = 0$$

$$-3x^5 - 4x^4 - 3x^3 + x + 1 = 0$$

The number of variations equals one, so the number of negative roots cannot exceed one.

Quiz: Equations of Higher Order

1. Find all the roots of the equation $(x + 1)(x^2 + 4x - 5) = 0$.

   (A) $\{-1, 4, -5\}$
   (B) $\{1, 2, 3\}$
   (C) $\{-5, -1, 1\}$
   (D) $\{1, 4, 5\}$
   (E) $\{-1, 2, 3\}$

2. Find the solutions of the equation $x^3 - 1 = 0$.

   (A) $\{-1, 1, 1\}$
   (B) $\{-1, -1, -1\}$
   (C) $\left\{1, \frac{-1+i\sqrt{3}}{2}, \frac{-1-i\sqrt{3}}{2}\right\}$
   (D) $\left\{1, \frac{-1+\sqrt{3}}{2}, \frac{-1-\sqrt{3}}{2}\right\}$
   (E) None of these.

3. Solve the equation $x^3 - 16x = 0$.

(A) $\{0, 4, 4\}$

(B) $\{0, -4, -4\}$

(C) $\{4, -4\}$

(D) $\{4, 4\}$

(E) $\{0, -4, 4\}$

4. Solve the equation $x^4 - 5x^2 - 36 = 0$.

(A) $\{2, -2, 3i, -3i\}$

(B) $\{2, 2, 3i, -3i\}$

(C) $\{2i, -2i, 3, -3\}$

(D) $\{2i, 2i, 3, 3\}$

(E) $\{2, -2, 3, -3\}$

5. Solve the equation $2x^4 + 7x^2 - 4 = 0$.

(A) $\pm\frac{\sqrt{2}}{2}, \pm 2i$

(B) $\pm\frac{2}{\sqrt{2}}, \pm 2i$

(C) $\frac{\pm\sqrt{2}}{2i}, \pm 2$

(D) $\frac{\pm\sqrt{2}}{2i}, \pm 2i$

(E) $\pm\frac{\sqrt{2}}{2}, \pm 2$

6. Solve for x: $x^4 - 10x^2 + 9 = 0$.

(A) $3, -1$

(B) $\pm 3, \pm 1$

(C) $3, 1$

(D) $-3, 1$

(E) $-3, -1$

7. Solve the equation $x^4 = 4x^2 - 4$.

(A) $\sqrt{2}, 2i$

(B) ± 4

(C) $\sqrt{2}, -2i$

(D) $\pm\sqrt{2}$

(E) ± 2

8. Solve the equation $(x^2 - 3x)^2 - 2(x^2 - 3x) - 8 = 0$.

(A) $\{4, 2, 1, -1\}$
(B) $\{-4, -2, 1, 1\}$
(C) $\{-4, -2, -1, -1\}$
(D) $\{4, -2, 1, -1\}$
(E) $\{-4, 2, 1, -1\}$

9. Solve for x: $x^4 + 6x^3 + 2x^2 - 21x - 18 = 0$.

(A) $1, 2, \frac{-3+3\sqrt{5}}{2}, \frac{-3+3\sqrt{5}}{2}$

(B) $1, 2, \frac{-3+3\sqrt{5}}{2}, \frac{3-3\sqrt{5}}{2}$

(C) $1, 2, \frac{-3+3\sqrt{5}}{2}, \frac{-3-3\sqrt{5}}{2}$

(D) $-1, -2, \frac{-3+3\sqrt{5}}{2}, \frac{3-3\sqrt{5}}{2}$

(E) $-1, -2, \frac{-3+3\sqrt{5}}{2}, \frac{-3-3\sqrt{5}}{2}$

10. Solve $x^4 - 2x^2 - 3 = 0$ as an equation in x^2.

(A) $\sqrt{3}, 1$
(B) $-\sqrt{3}, 1$
(C) $\pm\sqrt{3}, \pm i$
(D) $\sqrt{3}, i$
(E) $-\sqrt{3}, -i$

ANSWER KEY

1. (C)
2. (C)
3. (E)
4. (C)
5. (A)
6. (B)
7. (D)
8. (A)
9. (E)
10. (C)

CHAPTER 11

Ratio, Proportion, and Variation

11.1 Ratio and Proportion

The ratio of two numbers x and y written $x{:}y$ is the fraction $\frac{x}{y}$ where $y \neq 0$. A proportion is an equality of two ratios. The laws of proportion are listed below:

If $\frac{a}{b} = \frac{c}{d}$, then:

A) $ad = bc$

B) $\frac{b}{a} = \frac{d}{c}$

C) $\frac{a}{c} = \frac{b}{d}$

D) $\frac{a+b}{b} = \frac{c+d}{d}$

E) $\frac{a-b}{b} = \frac{c-d}{d}$

Given a proportion $a{:}b = c{:}d$, then a and d are called the extremes, b and c are called the means, and d is called the fourth proportional to a, b, and c.

Problem Solving Examples:

Solve the proportion $\frac{x+1}{4} = \frac{15}{12}$.

Cross multiply to determine x; that is, multiply the numerator of the first fraction by the denominator of the second, and equate this to the product of the numerator of the second and the denominator of the first.

$$\begin{aligned} (x+1)12 &= 4 \times 15 \\ 12x + 12 &= 60 \\ x &= 4 \end{aligned}$$

If $a/b = c/d$, $a + b = 60$, $c = 3$, and $d = 2$, find b.

We are given $\frac{a}{b} = \frac{c}{d}$. Cross multiplying, we obtain $ad = bc$. Adding bd to both sides, we have $ad + bd = bc + bd$, which is equivalent to $d(a + b) = b(c + d)$ or

$$\frac{a+b}{b} = \frac{c+d}{d}.$$

Replacing $(a + b)$ by 60, c by 3, and d by 2, we obtain

$$\frac{60}{b} = \frac{3+2}{2}$$

$$\frac{60}{b} = \frac{5}{2}$$

Cross multiplying, $5b = 120$

$$b = 24.$$

11.2 Variation

A) If x is directly proportional to y written $x \alpha y$, then $x = ky$ or $\frac{x}{y} = k$, where k is called the constant of proportionality or the constant of variation.

B) If x varies inversely as y, then $x = \frac{k}{y}$.

C) If x varies jointly as y and z, then $x = kyz$.

Example: If y varies jointly as x and z, and $3x{:}1 = y{:}z$, find the constant of variation.

Solution: A variable s is said to vary jointly as t and v if s varies directly as the product tv, that is, if $s = ktv$ where k is called the constant of variation.

Here the variable y varies jointly as x and z with k as the constant of variation.

$$y = kxz$$

$$3x{:}1 = y{:}z$$

Expressing these ratios as fractions,

$$\frac{3x}{1} = \frac{y}{z}$$

Solving for y by cross-multiplying,

$$y = 3xz$$

Equating both relations for y, we have:

$$kxz = 3xz$$

Solving for the constant of variation, k, we divide both sides by xz,

$$k = 3.$$

Problem Solving Examples:

If y varies directly with respect to x and $y = 3$ when $x = -2$, find y when $x = 8$.

If y varies directly as x, then y is equal to some constant k times x; that is, $y = kx$ where k is a constant. We can now say $y_1 = kx_1$ and $y_2 = kx_2$ or $\frac{y_1}{x_1} = k$, $\frac{y_2}{x_2} = k$ which implies $\frac{y_1}{x_1} = \frac{y_2}{x_2}$ which is a proportion. We use the proportion $\frac{y_1}{x_1} = \frac{y_2}{x_2}$. Thus $\frac{3}{-2} = \frac{y_2}{8}$. Now solve for y_2:

$$8\left(\frac{3}{-2}\right) = 8\left(\frac{y_2}{8}\right)$$

$$-12 = y_2.$$

When $x = 8$, $y = -12$.

If y varies inversely as the cube of x, and $y = 7$ when $x = 2$, express y as a function of x.

The relationship "y varies inversely with respect to x" is expressed as,

$$y = \frac{k}{x}.$$

The inverse variation is now with respect to the cube of x, x^3, and we have,

$$y = \frac{k}{x^3}$$

Since $y = 7$ and $x = 2$ must satisfy this relation, we replace x and y by these values,

$$7 = \frac{k}{2^3} = \frac{k}{8},$$

and we find $k = 7 \times 8 = 56$. Substitution of this value of k in the general relation gives,

$$y = \frac{56}{x^3},$$

which expresses y as a function of x. We may now, in addition, find the value of y corresponding to any value of x. If we had the added requirement to find the value of y when $x = 1.2$, $x = 1.2$ would be substituted in the function to give

$$y = \frac{56}{(1.2)^3} = \frac{56}{1.728} = 32.41.$$

Other expressions in use are "is proportional to" for "varies directly," and "is inversely proportional to" for "varies inversely."

CHAPTER 12

Logarithms

12.1 Logarithms

Logarithms were invented to make calculations with large numbers easier by expressing them in terms of their exponents. Given the exponential expression $x = b^y$, $b \in \mathbb{N}$, $b \neq 1$, and b is a positive real number, we can rewrite it in its logarithmic form so that x is in terms of its exponent y, $y = \log_b x$. This expression is read as "y is equal to the logarithm to the base b of x."

If x, y, and a are positive real numbers, $a \neq 1$, and r is any real number, then

A) $\log_a(xy) = \log_a x + \log_a y$

B) $\log_a\left(\frac{x}{y}\right) = \log_a x - \log_a y$

C) $\log_a x^r = r \log_a x$.

Common logarithms are logarithms with a base of 10. We omit writing the base when working with base 10. That is

$$\log x = \log_{10} x.$$

The following formula will enable us to calculate non-common logarithms by using common logarithms.

$$\log_a b = \frac{\log_{10} b}{\log_{10} a}, \text{ where } a, b > 0.$$

For example, to find the value of $\log_7 3$, we use the above rule to obtain:

$$\log_7 3 = \frac{\log_{10} 3}{\log_{10} 7} \cong \frac{.4771}{.8451} \cong .5645.$$

These values can be found by referring to log tables.

The antilogarithm is the number corresponding to a given logarithm. The cologarithm of a positive number is the logarithm of its reciprocal.

The common logarithm of any number is expressible as a combination of two parts:

A) the characteristic, which is the integral part; and

B) the mantissa, which is the decimal part of the number.

To find the common logarithm of a positive number:

A) Express the number in scientific notation.

B) Determine the index of the number, which is the characteristic.

C) To find the mantissa, see a table of common logarithms of numbers.

Example: Find the logarithm of 30,700.

Solution: First express 30,700 in scientific notation. $30{,}700 = 3.07 \times 10^4$. Four is the characteristic. To find the mantissa, see a table of common logarithms of numbers. The mantissa is .4871. Thus, $\log 30{,}700 = 4 + .4871 = 4.4871$.

To find the antilogarithm:

A) Use the logarithm table to find the number that corresponds to that specific mantissa.

B) Rewrite that number in standard form.

C) Use the characteristic as the index for the number in standard form.

Example: Find $\text{Antilog}_{10}\, 0.8762 - 2$.

Solution: Let $N = \text{Antilog}_{10}\, 0.8762 - 2$. The following relationship between log and antilog exists; $\log_{10} x = a$ is the equivalent of $x = \text{antilog}_{10} a$. Therefore,

$$\log_{10} N = 0.8762 - 2.$$

The characteristic is –2. The mantissa is 0.8762. The number that corresponds to this mantissa is 7.52. This number is found from a table of common logarithms, base 10. Therefore,

$$\begin{aligned} N &= 7.52 \times 10^{-2} \\ &= 7.52 \times \left(\frac{1}{10^2}\right) \\ &= 7.52 \times \left(\frac{1}{100}\right) \\ &= 7.52(.01) \\ N &= 0.0752. \end{aligned}$$

Therefore, $N = \text{Antilog}_{10} 0.8762 - 2 = 0.0752$.

Problem Solving Examples:

Write $5^3 = 125$ in logarithmic form.

The statement $b^y = x$ is equivalent to the statement $\log_b x = y$, where b is the base and y is the exponent. The latter form is the logarithmic form. Thus, $5^3 = 125$ in logarithmic form is $\log_5 125 = 3$, where the base is 5 and the logarithm of 125 is 3.

Find $\log_{10} 100$.

The following solution presents two methods for solving the given problem.

Method I. The statement $\log_{10} x = y$ is equivalent to $10^y = x$; hence, $\log_{10} 100 = x$ is equivalent to $10^x = 100$.

Since $10^2 = 100$, $\log_{10} 100 = 2$.

Method II. Note that $100 = 10 \times 10$; thus $\log_{10} 100 = \log_{10} (10 \times 10)$. Recall: $\log_x (a \times b) = \log_x a + \log_x b$, therefore

$$\begin{aligned}\log_{10} (10 \times 10) &= \log_{10} 10 + \log_{10} 10 \\ &= 1 + 1 \\ &= 2.\end{aligned}$$

12.2 Logarithmic, Exponential, and Power Functions

The function $f(x) = x^n$ is called a power function in x. An exponential function in x is of the form $f(x) = a^x$. A logarithmic function in x is of the form $f(x) = \log_a x$.

An equation involving one or more unknowns in an exponent is called an exponential equation. $2^x + 6x + 8^{y+7} = y$ is an exponential equation in two unknowns.

Figure 12.1 through 12.5 are examples of the graphs of power, exponential, and logarithmic functions.

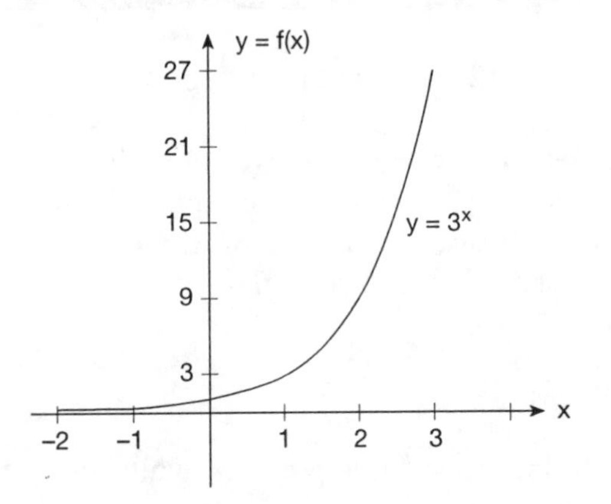

Figure 12.1 Exponential function

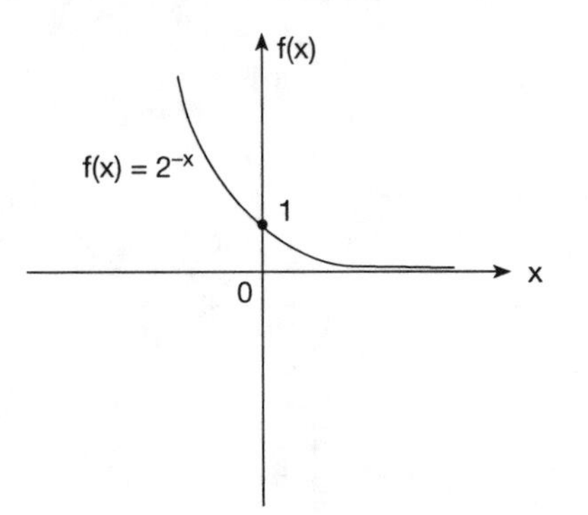

Figure 12.2 Exponential function

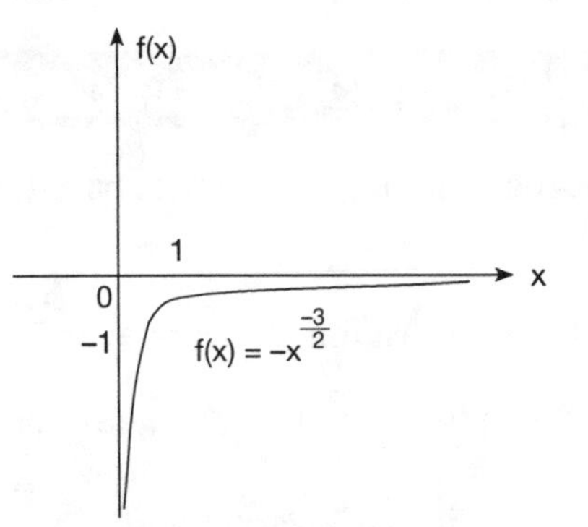

Figure 12.3 Power function

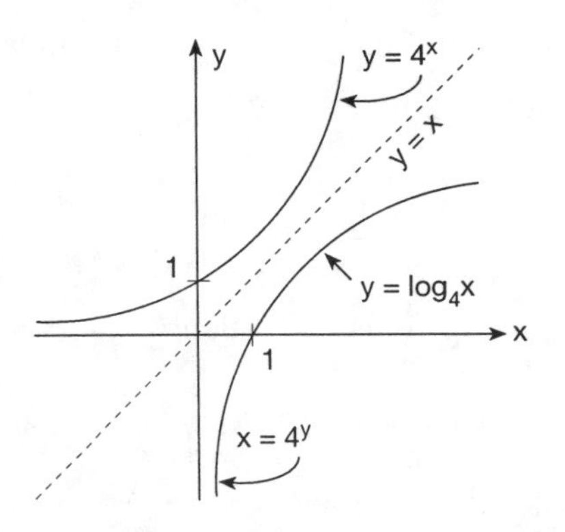

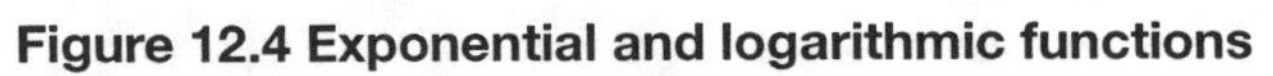
Figure 12.4 Exponential and logarithmic functions

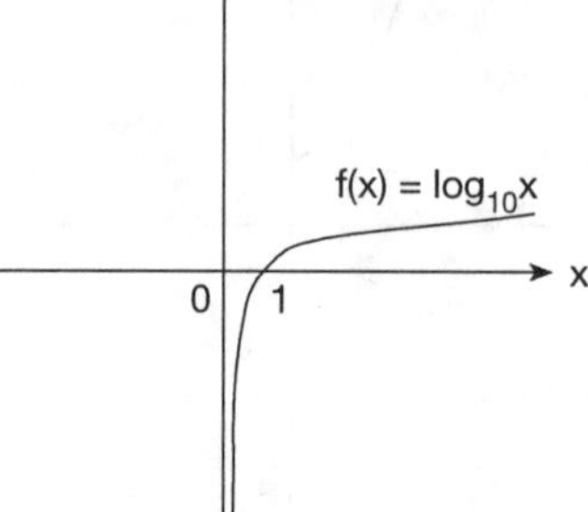

Figure 12.5 Logarithmic function

Problem Solving Examples:

If f is the logarithmic function with base 4, find $f(4)$, $f\left(\frac{1}{4}\right)$, and $f(8)$.

Since f is the logarithmic function with base 4, then $y = f(x) = \log_4 x$. The values $f(4)$, $f\left(\frac{1}{4}\right)$, and $f(8)$ can be found by replacing x by 4, $\frac{1}{4}$, and 8 in the logarithmic function $y = f(x) = \log_4 x$. Hence, $f(4) = \log_4 4$. Let $N_1 = f(4) = \log_4 4$. By definition, $\log_x a = N$ is equivalent to $x^N = a$. Therefore, $N_1 = \log_4 4$ is equivalent to $4^{N_1} = 4$. Since $4^1 = 4$, $N_1 = 1$. Then, $N_1 = 1 = f(4)$.

For the second value $f\left(\frac{1}{4}\right)$, $f\left(\frac{1}{4}\right) = \log_4 \frac{1}{4}$. Let $N_2 = f\left(\frac{1}{4}\right) = \log_4 \frac{1}{4}$. Hence, $N_2 = \log_4 \frac{1}{4}$ is equivalent to $4^{N_2} = \frac{1}{4}$. Since $4^{-1} = \frac{1}{4^1} = \frac{1}{4}$, $N_2 = -1$. Then, $N_2 = -1 = f\left(\frac{1}{4}\right)$.

For the third value $f(8)$, $f(8) = \log_4 8$. Let $N_3 = f(8) = \log_4 8$. Hence, $N_3 = \log_4 8$ is equivalent to $4^{N_3} = 8$. Since $4 = 2^2$, $N_3 = \log_4 8$ is equivalent to $\left(2^2\right)^{N_3} = 8$ or $2^{2N_3} = 8$. Since $2^3 = 8$, $2N_3 = 3$. Dividing both sides of the equation $2N_3 = 3$ by 2:

$$\frac{2N_3}{2} = \frac{3}{2} \qquad \text{or} \qquad N_3 = \frac{3}{2}$$

Then, $N_3 = \dfrac{3}{2} = f(8)$.

Q Solve the "exponential" equation: $2^{0.4x} = 7$.

A 
Taking the log of both sides of the given equation,

$$\log 2^{0.4x} = \log 7.$$

Since $\log_b y^a = a \log_b y$, $\log 2^{0.4x} = 0.4x \log 2$.

Thus, $\qquad 0.4x \log 2 = \log 7$

or $\qquad x = \dfrac{\log 7}{0.4 \log 2}$.

In a log table we find $\log 7 = .8451$ and $\log 2 = .3010$. Thus,

$$x = \frac{.8451}{(0.4)(.3010)} = \frac{.8451}{.1204} = 7.02.$$

Or, if we wish, we may complete the solution by using logarithms again:

$$\log 0.8451 = \log 8.451 \times 10^{-1} = .9269 - 1$$

$$\log 0.1204 = \log 1.204 \times 10^{-1} = .0806 - 1.$$

Since $\log_b \dfrac{y}{z} = \log_b y - \log_b z$,

$$\log x = \log \frac{.8451}{.1204} = \log .8451 - \log .1204$$

$$= (.9269 - 1) - (.0806 - 1).$$

Thus, $\log x = .8463.$

We look up the mantissa, 0.8463, in a table of common logarithms and find its corresponding number to be 7020. We adjust the decimal point by noting the characteristic 0 of 0.8463 is one less than the number of digits to the left of the decimal point of the number we seek. In this case, therefore, there should be one digit to the left of the decimal point. Hence,

$$x = 7.020.$$

CHAPTER 13

Sequences and Series

13.1 Sequences

A set of numbers u_1, u_2, u_3, ... in a definite order of arrangement and formed according to a definite rule is called a sequence. Each number in the sequence is called a term of the sequence. If the number of terms is finite, it is called a finite sequence, otherwise, it is called an infinite sequence.

A general term n can be obtained by applying a general law of formation, by which any term in the sequence can be obtained.

An arithmetic progression (AP) is a sequence of numbers where each term, excluding the first, is obtained from the preceding one by adding a fixed quantity to it. This constant amount is called the common difference.

Let $a =$ first term of progression

$a_n =$ nth term of progression

$l =$ last term

$d =$ common difference

$k =$ number of terms

$S_n =$ sum of first n terms, $n \leq k$

then $$l = a + (k - 1)d$$

$$S_n = \frac{n}{2}(a + a_n) = \frac{n}{2}[2a + (n - 1)d].$$

In general, to find the common difference in a given arithmetic progression, simply subtract any term from its successor.

A single term between two given terms a and b is called the arithmetic mean, M, between these two terms and is given by:

$$M = \frac{a+b}{2}.$$

Example: Insert four arithmetic means between 1 and 36.

Solution: The terms between any two given terms of a progression are called the means between these two terms. Inserting four arithmetic means between 1 and 36 requires an arithmetic progression (A.P.) of the form 1, ___, ___, ___, ___, 36, where the second term is the mean of the first and third terms, the third term is the mean of the second and fourth terms, etc. Using the formula, $l = a + (k - 1)d$, for the nth term, we can determine d. Knowing the common difference, d, we can obtain the means by adding d to each preceding number after the first.

$$\begin{aligned} a &= 1, l = 36, \text{ and } k = 6 \text{ since there will be six terms} \\ l &= a + (k - 1)d \\ 36 &= 1 + 5d \\ 5d &= 35 \\ d &= 7 \end{aligned}$$

The arithmetic means are: $1 + 7$, $(1 + 7) + 7$, $(1 + 7 + 7) + 7$, $(1 + 7 + 7 + 7) + 7$; that is, 8, 15, 22, and 29. The arithmetic progression is 1, 8, 15, 22, 29, 36.

A geometric progression (GP) is a sequence of numbers each of which, after the first, is obtained by multiplying the preceding number by a constant number called the common ratio, r.

Let a = first term

a_n = nth term of progression

r = common ratio

k = number of terms

l = last term

S_n = sum of first n terms

then

$$l = ar^{k-1}$$

$$S_n = \frac{a(r^k - 1)}{r - 1} = \frac{rl - a}{r - 1}.$$

The sum to infinity (S_∞) of any geometric progression is

$$S_\infty = \frac{a}{l - r} \quad \text{if} \quad |r| < 1.$$

To find the common ratio in a given geometric progression, divide any term by its predecessor.

A single term between two given terms of a geometric progression is called the geometric mean between the two terms. The geometric mean is denoted by G.

If G is the geometric mean between the terms a and b, then $\frac{a}{G} = \frac{G}{b}$ which implies:

$$G^2 = ab \quad \text{or} \quad G = \pm\sqrt{ab}.$$

A harmonic progression (HP) is a sequence of numbers whose reciprocals are an arithmetic progression. The terms between any two given terms of a harmonic progression are called the harmonic means between these two terms.

Let a = first term

a_n = nth term of progression

l = last term

k = number of terms

d = common difference

then
$$l = \frac{1}{a+(k-1)d}.$$

Example: If a^2, b^2, and c^2 are in arithmetic progression, show that $b + c$, $c + a$, and $a + b$ are in harmonic progression.

Solution: We are given that a^2, b^2, and c^2 are in arithmetic progression. By this we mean that each new term is obtained by adding a constant to the preceding term.

By adding $(ab + ac + bc)$ to each term, we see that

$$a^2 + (ab + ac + bc),\ b^2 + (ab + ac + bc),\ c^2 + (ab + ac + bc)$$

are also in arithmetic progression. These three terms can be rewritten as

$$a^2 + ab + ac + bc,\ b^2 + bc + ab + ac,\ c^2 + ac + bc + ab.$$

Notice that:

$$\begin{aligned} a^2 + ab + ac + bc &= (a + b)(a + c) \\ b^2 + bc + ab + ac &= (b + c)(b + a) \\ c^2 + ac + bc + ab &= (c + a)(c + b) \end{aligned}$$

Therefore, the three terms can be rewritten as:

$$(a + b)(a + c),\ (b + c)(b + a),\ (c + a)(c + b)$$

which are also in arithmetic progression.

Now, dividing each term by $(a+b)(b+c)(c+a)$, we obtain:

$$\frac{1}{b+c}, \quad \frac{1}{c+a}, \quad \frac{1}{a+b}$$

which are in arithmetic progression.

Recall that a sequence of numbers whose reciprocals form an arithmetic progression is called a harmonic progression. Thus, since $\frac{1}{b+c}$, $\frac{1}{c+a}$, $\frac{1}{a+b}$ is an arithmetic progression, $b+c$, $c+a$, and $a+b$ are in harmonic progression.

Problem Solving Examples:

If the first term of a geometric progression is 9 and the common ratio is $-\frac{2}{3}$, find the first five terms.

A geometric progression (GP) is a sequence of numbers each of which, after the first, is obtained by multiplying the preceding number by a constant number called the common ratio, r. Thus, a GP such as $a_1, a_2, a_3, a_4, a_5, \ldots$ or $a_1, a_1r, a_2r, a_3r, a_4r, \ldots$ with $a_1 = 9$ and $r = -\frac{2}{3}$ is determined as follows:

$$a_1 = 9$$

$$a_2 = 9\left(-\frac{2}{3}\right) = -6$$

$$a_3 = (-6)\left(-\frac{2}{3}\right) = 4$$

$$a_4 = 4\left(-\frac{2}{3}\right) = -\frac{8}{3}$$

$$a_5 = \left(-\frac{8}{3}\right)\left(-\frac{2}{3}\right) = \frac{16}{9}$$

Thus, the first five terms are: $9, -6, 4, -\frac{8}{3}, \frac{16}{9}$.

Find the sum of the first ten terms of the geometric progression: 15, 30, 60, 120, …

A geometric progression is a sequence in which each term after the first is formed by multiplying the preceding term by a fixed number, called the common ratio.

If a is the first term, r is the common ratio, and n is the number of terms, the geometric progression (G.P.) is

$$a, ar, ar^2, \ldots, ar^{n-1}$$

The given GP, 15, 30, 60, 120, …, may be written as 15, 15(2), 15(2^2), 15(2^3) … . The sum, S_n, of the first n terms of the geometric progression is given by

$$S_n = \frac{a(r^n - 1)}{r - 1}$$

where a = first term

r = common ratio

n = number of terms.

Here $a = 15$, $r = 2$, and $n = 10$.

$$S_{10} = \frac{15(2^{10} - 1)}{2 - 1}$$

$$= \frac{15(1{,}024 - 1)}{1}$$

$$= 15(1{,}023)$$

$$= 15{,}345$$

13.2 Series

A series is defined as the sum of the terms of a sequence $u_1 + u_2 + u_3 + \ldots + u_n \ldots$

The terms $u_1, u_2, u_3, \ldots, u_n$ are called the first, second, third, and nth terms of the series. If the series has a finite number of terms, it is called a finite series, otherwise, it is called an infinite series.

Finite series

$$\sum_{i=1}^{n} u_i = u_1 + u_2 + u_3 + \ldots + u_n$$

Infinite series

$$\sum_{i=1}^{\infty} u_i = u_1 + u_2 + u_3 + \ldots$$

The general, or nth, term of a series is an expression which indicates the law of formation of the terms.

Example: Determine the general term of the series:

$$\frac{1}{5^3} + \frac{3}{5^5} + \frac{5}{5^7} + \frac{7}{5^9} + \frac{9}{5^{11}} + \ldots$$

Solution: The numerators of the terms in the series are consecutive odd numbers beginning with 1. An odd number can be represented by $2n - 1$.

In the denominators, the base is always 5, and the power is a consecutive odd integer beginning with 3.

The general term can therefore be expressed by

$$\frac{2n-1}{5^{2n+1}}, \text{ the series by } \sum_{i=1}^{\infty} \frac{2i-1}{5^{2i+1}}$$

and the series is generated by replacing i with $i = 1, 2, 3, 4, \ldots$.

Let $s_n = u_1 + u_2 + \ldots + u_n$ be the sum of the first n terms of the infinite series $u_1 + u_2 + u_3 + \ldots$. The terms of the sequence $s_1, s_2, s_3 \ldots$ are called the partial sums of the series.

If the values of $s_1, s_2, \ldots, s_n$ never become greater than or equal to a certain value S, no matter how big n is, and approach S as n increases, the sums are said to have a limit. This is represented by:

$$\lim_{n \to \infty} s_n = S.$$

If $\lim_{n \to \infty} s_n = S$ is a finite number, the series $u_1 + u_2 + u_3 + \ldots$ is convergent and S is called the sum of the infinite series.

A series which is not convergent is said to be divergent.

An alternating series is one whose terms are alternately positive and negative. An alternating series converges if:

A) After a certain number of terms the absolute value of a certain term is less than that of the preceding term.

B) The nth term has a limit of zero as $n \to \infty$.

A series is said to be absolutely convergent if the series formed by taking absolute values of the terms converges. A convergent series which is not absolutely convergent is conditionally convergent.

The terms of an absolutely convergent series may be arranged in any order and not affect the convergence.

A series of the form $c_0 + c_1 x + c_2 x^2 + \ldots$ where the coefficients $c_0, c_1, c_2, \ldots$ are called constants is called a power series in x. It is denoted by

$$\sum_{n=0}^{\infty} c_n x^n .$$

The set of values of x for which a power series converges is called its interval of convergence.

A series of the type $\frac{1}{1^p} + \frac{1}{2^p} + \frac{1}{3^p}$ where p is a constant is known as a p series and is denoted by

$$\sum_{n=1}^{\infty} \frac{1}{n^p} .$$

The p series converges if $p \leq 1$.

The following methods are used to test convergence of series:

A) Comparison test for convergence of series of positive terms.

If each term of a given series of positive terms is less than or equal to the corresponding term of a known convergent series, then the given series converges.

If each term of a given series of positive terms is greater than or equal to the corresponding term of a known divergent series, then the given series diverges.

Example: Establish the convergence or divergence of the series:

$$\frac{1}{1+\sqrt{1}} + \frac{1}{1+\sqrt{2}} + \frac{1}{1+\sqrt{3}} + \frac{1}{1+\sqrt{4}} + \ldots.$$

Solution: To establish the convergence or divergence of the given series, we first determine the nth term of the series. By studying the law of formation of the terms of the series, we find the nth term to be $\frac{1}{1+\sqrt{n}}$.

To determine whether this series is convergent or divergent, we use the comparison test. We choose $\frac{1}{n}$, which is a known divergent series since it is a p-series, $\frac{1}{n^p}$, with $p = 1$. If we can show $\frac{1}{1+\sqrt{n}} > \frac{1}{n}$, then $\frac{1}{1+\sqrt{n}}$ is divergent. But we can see this is true, since $1 + \sqrt{n} < n$ for $n > 2$. Therefore, the given series is divergent.

B) Ratio test:

For a given series $s_1 + s_2 + s_3 + \ldots$, it is possible to conclude it is:

Convergent if: $\lim_{n\to\infty} \left| \frac{s_{n+1}}{s_n} \right| = L < 1$

Divergent if: $\lim_{n\to\infty} \left| \frac{s_{n+1}}{s_n} \right| = L > 1$

If $\lim_{n\to\infty}\left|\frac{s_{n+1}}{s_n}\right| = L = 1$, the ratio test is not decisive; it fails to establish convergence or divergence.

Problem Solving Examples:

Q Find the numerical value of the following:

a) $\sum_{j=1}^{7}(2j+1)$ b) $\sum_{j=1}^{21}(3j-2)$

If $A(r)$ is some mathematical expression and n is a positive integer, then the symbol $\sum_{r=0}^{n} A(r)$ means "Successively replace" the letter r in the expression $A(r)$ with the numbers 0, 1, 2, …, n and add up the terms. The symbol Σ is the Greek letter sigma and is a shorthand way to denote "the sum." It avoids having to write the sum $A(0) + A(1) + A(2) + \dots + A(n)$.

A) For A) successively replace j by 1, …,7 and add up the terms.

$$\begin{aligned}\sum_{j=1}^{7}(2j+1) &= (2(1)+1)+(2(2)+1)+(2(3)+1)+(2(4)+1)+\\ &\quad (2(5)+1)+(2(6)+1)+(2(7)+1)\\ &= (2+1)+(4+1)+(6+1)+(8+1)+(10+1)+\\ &\quad (12+1)+(14+1)\\ &= 3+5+7+9+11+13+15\\ &= 63\end{aligned}$$

B) For B) successively replace j by 1, 2, 3, …, 21 and add up the terms.

$$\begin{aligned}\sum_{j=1}^{21}(3j-2) &= (3(1)-2)+(3(2)-2)+(3(3)-2)+(3(4)-2)+\\ &\quad (3(5)-2)+(3(6)-2)+(3(7)-2)+(3(8)-2)+\\ &\quad (3(9)-2)+(3(10)-2)+(3(11)-2)+(3(12)-2)+\\ &\quad (3(13)-2)+(3(14)-2)+(3(15)-2)+(3(16)-2)+\\ &\quad (3(17)-2)+(3(18)-2)+(3(19)-2)+(3(20)-2)+\\ &\quad (3(21)-2)\\ &= (3-2)+(6-2)+(9-2)+(12-2)+(15-2)+(18-2)+\\ &\quad (21-2)+(24-2)+(27-2)+(30-2)+(33-2)+\end{aligned}$$

$$\begin{aligned} & (36-2) + (39-2) + (42-2) + (45-2) + (48-2) + \\ & (51-2) + (54-2) + (57-2) + (60-2) + (63-2) \\ = \; & 1 + 4 + 7 + 10 + 13 + 16 + 19 + 22 + 25 + 28 + 31 + \\ & 34 + 37 + 40 + 43 + 46 + 49 + 52 + 55 + 58 + 61 \\ = \; & 651 \end{aligned}$$

Determine the general term of the sequence:

$$\frac{1}{2}, \frac{1}{12}, \frac{1}{30}, \frac{1}{56}, \frac{1}{90}, \ldots$$

To determine the general term, it is necessary to find how the adjacent terms differ. In this example, it is sufficient to consider the denominator because the numerator is the same for all the terms.

Now we try to write an expression that generates the series. By inspection, each term is the product of two successive integers, for example:

$$\frac{1}{2} = \frac{1}{1} \times \frac{1}{2}, \quad \frac{1}{12} = \frac{1}{3} \times \frac{1}{4}, \quad \frac{1}{30} = \frac{1}{5} \times \frac{1}{6}, \quad \frac{1}{56} = \frac{1}{7} \times \frac{1}{8}$$

This fact can be expressed as

$$\frac{1}{(2n-1)(2n)}$$

and this is the desired answer.

Quiz: Ratio, Proportion, and Variation—Sequences and Series

1. If $\log_8 x = \frac{4}{3}$, then $x =$

(A) 1.

(B) 2.

(C) 4.

(D) 8.

(E) 16.

2. $\log_2 \dfrac{\sqrt{2}}{8} =$

(A) $\dfrac{5}{2}$

(B) $\dfrac{1}{2}$

(C) $\dfrac{3}{2}$

(D) $-\dfrac{5}{2}$

(E) $-\dfrac{1}{2}$

3. For the following sequence of numbers, $\dfrac{1}{2}, \dfrac{1}{12}, \dfrac{1}{30}, \ldots$, the next number will be

(A) $\dfrac{1}{36}$.

(B) $\dfrac{1}{27}$.

(C) $\dfrac{1}{48}$.

(D) $\dfrac{1}{56}$.

(E) $\dfrac{1}{72}$.

4. The first three terms of a progression are 3, 6, 12, …. What is the value of the tenth term?

(A) 1,200

(B) 2,468

(C) 188

(D) 1,536

(E) 3,272

5. If the sum of three consecutive odd numbers is 51, then the first odd number of the sequence is

(A) 11.

(B) 13.

(C) 15.

(D) 17.

(E) 19.

6. $\log_4 64$ is identical to

(A) $\log_7 343$.
(B) $\dfrac{\log_{10} 64}{\log_{10} 4}$.
(C) $\log_8 256$.
(D) Both (A) and (B).
(E) Both (A) and (C).

7. If $\log_{10} 3 = .4771$ and $\log_{10} 4 = .6021$, find $\log_{10} 12$.

(A) 1.262
(B) 1.9084
(C) 1.0792
(D) 1.8063
(E) 0.2873

8. Evaluate $\dfrac{542.3\sqrt{0.1383}}{32.72}$ using logarithms.

(A) 1.5148
(B) 2.7343
(C) –1.1408
(D) 6.163
(E) 0.7898

9. Find the sum of the first six terms of a geometric progression whose first term is 1/3 and whose second term is –1.

(A) $\dfrac{182}{3}$
(B) $-\dfrac{365}{6}$
(C) $-\dfrac{364}{3}$
(D) $-\dfrac{365}{3}$
(E) $-\dfrac{182}{3}$

10. Find the ninth term of the harmonic progression 3, 2, $\frac{3}{2}$, ...

(A) $\frac{11}{6}$.
(B) $\frac{3}{5}$.
(C) $\frac{6}{11}$.
(D) $\frac{5}{3}$.
(E) $\frac{2}{3}$.

ANSWER KEY

1. (E)
2. (D)
3. (D)
4. (D)
5. (C)
6. (D)
7. (C)
8. (D)
9. (E)
10. (B)

CHAPTER 14

Permutations, Combinations, and Probability

14.1 Permutation

A permutation is an arrangement of all or part of a number of objects in any order.

The symbol ${}_bP_a$ represents the number of permutations of "b" objects taken "a" at a time.

$${}_bP_a = \frac{b!}{(b-a)!}$$

Example: $${}_9P_4 = \frac{9!}{(9-4)!} = \frac{9!}{5!} = \frac{9 \times 8 \times 7 \times 6 \times 5 \times 4 \times 3 \times 2 \times 1}{5 \times 4 \times 3 \times 2 \times 1}$$

$$= 3{,}024$$

Problem Solving Examples:

Calculate the number of permutations of the letters a, b, c, and d taken two at a time.

The first of the two letters may be taken in four ways (a, b, c, or d). The second letter may therefore be selected from the remaining three letters in three ways. By the fundamental counting principle the total number of ways of selecting two letters is equal to the product of the number of ways of selecting each letter; hence,

$$p(4, 2) = 4 \times 3 = 12.$$

The list of these permutations is:

ab	*ba*	*ca*	*da*
ac	*bc*	*cb*	*db*
ad	*bd*	*cd*	*dc*

In how many ways may three books be placed next to each other on a shelf?

We construct a pattern of three boxes to represent the places where the three books are to be placed next to each other on the shelf:

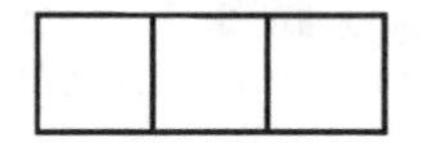

Since there are three books, the first place may be filled in three ways. There are then two books left, so that the second place may be filled in two ways. There is only one book left to fill the last place. Hence, our boxes take the following form:

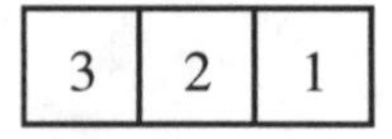

The Fundamental Principle of Counting states that if one thing can be done in a different ways and, when it is done in any one of these ways, a second thing can be done in b different ways, and a third thing can be done in c ways, … then all the things in succession can be done in $a \times b \times c$ … different ways. Thus, the books can be arranged in $3 \times 2 \times 1 = 6$ ways.

This can also be seen using the following approach. Since the arrangement of books on the shelf is important, this is a permutations problem. Recalling the general formula for the number of permutations of n things taken r at a time, ${}_nP_r = n!/(n-r)!$, we replace n by 3 and r by 3 to obtain

$${}_3P_3 = \frac{3!}{(3-3)!} = \frac{3!}{0!} = \frac{3\times2\times1}{1} = 6.$$

14.2 Combination

A combination is a grouping of all or part of a number of objects without regard to the order of the arrangement of the selected objects.

The symbol ${}_bC_a$, represents the combination of "b" different objects taken "a" at a time without regard to the order of the arrangement of the selected objects.

$${}_bC_a = \frac{b!}{a!(b-a)!} = \frac{{}_bP_a}{a!}$$

Note that ${}_bC_a$ can also be written as $\binom{b}{a}$, read "b choose a."

Example: ${}_9C_4 = \dfrac{{}_9P_4}{4!} = \dfrac{3,024}{24} = 126$

Also note that ${}_bC_a = {}_bC_{b-a}$.

Given n objects, if we take them one at a time, two at a time, three at a time, …, n at a time and add all these combinations, we will obtain C which is

$$C = 2^n - 1.$$

Problem Solving Examples:

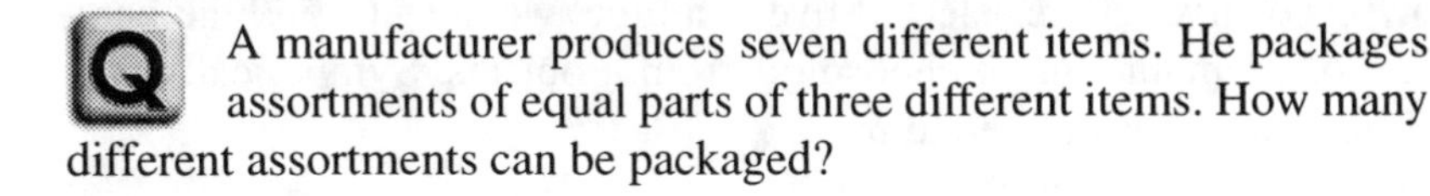

Q A manufacturer produces seven different items. He packages assortments of equal parts of three different items. How many different assortments can be packaged?

Since we are not concerned with the order of the items, we are dealing with combinations. Thus, the number of assortments is the number of combinations of seven items taken three at a time. Recall the general formula for the number of combinations of n items taken r at a time,

$$
\begin{aligned}
C(n, r) &= \frac{n!}{r!(n-r)!} \\
C(7, 3) &= \frac{7!}{3!(7-3)!} \\
&= \frac{7!}{3! \times 4!} \\
&= \frac{7 \times \cancel{6} \times 5 \times \cancel{4!}}{\cancel{3} \times \cancel{2} \times \cancel{4!}} \\
&= 35
\end{aligned}
$$

Thus, 35 different assortments can be packaged.

How many "words," each consisting of two vowels and three consonants, can be formed from the letters of the word "integral"?

To find the number of ways to choose vowels or consonants from letters, we use combinations. The number of combinations of n different objects taken r at a time is defined to be

$$C(n, r) = \frac{n!}{r!(n-r)!}.$$

Then, we first select the two vowels to be used, from among the three vowels in "integral"; this can be done in $C(3, 2) = 3$ ways. Next, we select the three consonants from the five in "integral"; this yields $C(5, 3) = 10$ possible choices. To find the number of ordered arrangements of five letters selected five at a time, we need to find the number of permutations of choosing r from n objects. Symbolically, it is $P(n, r)$ which is defined to be

$$P(n, r) = \frac{n!}{(n-r)!}.$$

We permute the five chosen letters in all possible ways, of which there are $P(5, 5) = 5! = 120$ arrangements. Finally, to find the total number of words which can be formed, we apply the Fundamental Counting Principle which states that if one event can be performed in m ways, another one in n ways, and another in k ways, then the total number of ways in which all events can occur is $m \times n \times k$ ways. Hence the total number of possible words is, by the fundamental principle,

$$C(3, 2)C(5, 3)P(5, 5) = 3 \times 10 \times 120 = 3{,}600.$$

14.3 Probability

If an event can occur in k ways and fail to occur in m ways and all of these $(k + m)$ ways are assumed equally likely, then the probability of the event to occur (success) is $P(s) = \frac{k}{k+m}$ and the probability of the event not to occur (failure) is $P(f) = \frac{m}{k+m}$.

14.3.1 Law of Total Probability

$$P(s) + P(f) = 1$$

This law states that given an event it is either a success or a failure.

Two or more events are said to be independent if the occurrence of any of them does not affect the probability of occurrence of any of the others.

The probability that two or more independent events will occur one after the other equals the product of their separate probabilities.

Two or more events are said to be dependent if the occurrence of one of the events affects the probability of occurrence of the others.

Two or more events are said to be mutually exclusive if the occurrence of any of them excludes the occurrence of the others.

The probability of occurrence of one, two, or more mutually exclusive events is the sum of the probabilities of the individual events.

Let p be the probability that a given event will occur in any single trial and q be the probability that it will fail to occur in any single trial. Then the probability that the event will occur exactly k times in n independent trials is ${}_nC_k\, p^k q^{n-k}$.

The probability that a given event will occur *at least m* times in n independent trials is given below:

$$P(s \geq m) = {}_nC_m\, p^m q^{n-m} + {}_nC_{m+1}\, p^{m+1} q^{n-(m+1)} + {}_nC_{m+2} p^{m+2} q^{n-(m+2)} + \ldots + {}_nC_n\, p^n q^o$$

where $q = 1 - p$.

Note that the probability that an event will occur at least m times in n trials corresponds to the sum of the probabilities of the event to occur m, $m + 1$, $m + 2$, …, n times.

The probability of any event to occur is between zero and one inclusive.

$$0 \leq P(s) \leq 1$$

$P(s) = 0$ indicates the event will not occur.

$P(s) = 1$ indicates the event will definitely occur.

Example: A die is tossed five times. What is the probability that a one will appear at least twice?

To find the probability that a one will occur at least twice, find the probability that it will occur twice, three times, four times, and five times. The sum of these probabilities will be that a one will happen at least twice. $P(s)$ = probability that a one will occur in a given trial.

$$P(s) = \frac{\text{number of ways to obtain a one}}{\text{number of ways to obtain any face of a die}}$$

An experiment can only succeed or fail; hence, the probability of success, $p(s)$, plus the probability of failure, $p(f)$, is one; $p(s) + p(f) = 1$.

Then $p(f) = 1 - p(s) = 1 - \frac{1}{6} = \frac{5}{6}$.

$$P(s \geq 2) = {}_5C_2\left(\frac{1}{6}\right)^2\left(\frac{5}{6}\right)^3 + {}_5C_3\left(\frac{1}{6}\right)^3\left(\frac{5}{6}\right)^2 + {}_5C_4\left(\frac{1}{6}\right)^4\left(\frac{5}{6}\right)^1 + {}_5C_5\left(\frac{1}{6}\right)^5\left(\frac{5}{6}\right)^0.$$

Then,

$${}_5C_2\left(\frac{1}{6}\right)^2\left(\frac{5}{6}\right)^3 + {}_5C_3\left(\frac{1}{6}\right)^3\left(\frac{5}{6}\right)^2 + {}_5C_4\left(\frac{1}{6}\right)^4\left(\frac{5}{6}\right)^1 + {}_5C_5\left(\frac{1}{6}\right)^5\left(\frac{5}{6}\right)^0$$

$$= \frac{5!}{2!3!}\left(\frac{125}{6^5}\right) + \frac{5!}{2!3!}\left(\frac{25}{6^5}\right) + \frac{5!}{4!1!}\left(\frac{5}{6^5}\right) + \frac{5!}{5!0!}\left(\frac{1}{6^5}\right)$$

$$= \frac{5 \times 4 \times 3!}{2 \times 3!}\left(\frac{125}{6^5}\right) + \frac{5 \times 4 \times 3!}{2 \times 3!}\left(\frac{25}{6^5}\right) + \frac{5 \times 4!}{4!1!}\left(\frac{5}{6^5}\right) + \frac{1}{6^5}$$

$$= 10\left(\frac{125}{6^5}\right) + 10\left(\frac{25}{6^5}\right) + 5\left(\frac{5}{6^5}\right) + \frac{1}{6^5}$$

$$= \frac{1,250 + 250 + 25 + 1}{6^5} = \frac{1,526}{7,776} = \frac{763}{3,888}.$$

Therefore, the probability that a one will appear at least twice is $\frac{763}{3,888}$.

Problem Solving Examples:

A deck of playing cards is thoroughly shuffled and a card is drawn from the deck. What is the probability that the card drawn is the ace of diamonds?

The probability of an event occurring is

$$\frac{\text{the number of ways the event can occur}}{\text{the number of possible outcomes}}.$$

In our case there is one way the event can occur, for there is only one ace of diamonds and there are 52 possible outcomes (for there are 52 cards in the deck). Hence, the probability that the card drawn is the ace of diamonds is 1/52.

A bag contains four black and five blue marbles. A marble is drawn and then replaced, after which a second marble is drawn. What is the probability that the first is black and the second blue?

Let C = event that the first marble drawn is black.
D = event that the second marble drawn is blue.

The probability that the first is black and the second is blue can be expressed symbolically:

$$P(C \text{ and } D) = P(CD).$$

We can apply the following theorem. If two events, A and B, are independent, then the probability that A and B will occur is

$$P(A \text{ and } B) = P(AB) = P(A) \times P(B).$$

Note that two or more events are said to be independent if the occurrence of one event has no effect upon the occurrence or non-occurrence of the other. In this case the occurrence of choosing a black marble has no effect on the selection of a blue marble and vice versa; since, when a marble is drawn it is then replaced before the next marble is drawn. Therefore, C and D are two independent events.

$$P(CD) = P(C) \times P(D)$$

$$P(C) = \frac{\text{number of ways to choose a black marble}}{\text{number of ways to choose a marble}}$$

$$= \frac{4}{9}$$

$$P(D) = \frac{\text{number of ways to choose a blue marble}}{\text{number of ways to choose a marble}}$$

$$= \frac{5}{9}$$

$$P(CD) = P(C) \times P(D) = \frac{4}{9} \times \frac{5}{9} = \frac{20}{81}$$

Q A traffic count at a highway junction revealed that out of 5,000 cars that passed through the junction in one week, 3,000 turned to the right. Find the probability that a car will turn (A) to the right and (B) to the left. Assume that the cars cannot go straight or turn around.

A (A) If an event can happen in s ways and fail to happen in f ways, and if all these ways $(s + f)$ are assumed to be equally likely, then the probability (p) that the event will happen is

$$p = \frac{s}{s+f} = \frac{\text{successful ways}}{\text{total ways}}.$$

In this case $s = 3{,}000$ and $s + f = 5{,}000$. Hence, $p = \frac{3{,}000}{5{,}000} = \frac{3}{5}$.

(B) If the probability that an event will happen is $\frac{a}{b}$, then the probability that this event will not happen is $1 - \frac{a}{b}$. Thus, the probability that a car will not turn right, but left, is $1 - \frac{3}{5} = \frac{2}{5}$. This same conclusion can also be arrived at using the following reasoning:

Since 3,000 cars turned to the right, 5,000 – 3,000 = 2,000 cars turned to the left. Hence, the probability that a car will turn to the left is

$$\frac{2{,}000}{5{,}000} = \frac{2}{5}.$$

Quiz: Permutations, Combinations, and Probability

1. How many games would it take a baseball coach to try every possible batting order with his nine players?

   (A) 9
   (B) 45
   (C) 81
   (D) 362,880
   (E) 3.8742×10^8

2. What is the probability of drawing an ace from a well-shuffled deck of 52 cards?

   (A) 0.0769
   (B) 0.0192
   (C) 0.0196
   (D) 0.0385
   (E) 0.5000

3. In how many different ways can the letters *a, b, c,* and *d* be arranged if they are selected three at a time?

   (A) 8
   (B) 12
   (C) 24
   (D) 4
   (E) 48

4. What is the probability that in a single throw of two dice the sum of 10 will appear?

(A) $\frac{10}{36}$

(B) $\frac{1}{6}$

(C) $\frac{1}{12}$

(D) $\frac{2}{10}$

(E) $\frac{11}{12}$

5. A bag contains four white balls, six black balls, three red balls, and eight green balls. If one ball is drawn from the bag, find the probability that it will be either white or green.

   (A) $\frac{1}{3}$

   (B) $\frac{2}{3}$

   (C) $\frac{4}{7}$

   (D) $\frac{4}{13}$

   (E) $\frac{8}{21}$

6. A box contains 30 blue balls, 40 green balls, and 15 red balls. What is the probability of choosing a red ball first followed by a blue ball?

   (A) 0.3571

   (B) 0.1765

   (C) 0.3529

   (D) 0.0620

   (E) 0.0630

7. When rolling a six-sided die, what is the probability of getting either a four or five?

   (A) 0.5000

   (B) 0.3333

   (C) 0.2500

   (D) 0.1670

   (E) 3.0000

8. What is the probability of getting at most one head in three coin tosses?

(A) 0

(B) $\frac{1}{4}$

(C) $\frac{1}{2}$

(D) $\frac{3}{4}$

(E) $\frac{7}{8}$

9. In how many ways can we arrange four letters (*a, b, c,* and *d*) in different orders?

(A) 4

(B) 8

(C) 16

(D) 24

(E) 48

10. Six dice are thrown. What is the probability of getting six ones?

(A) 0.0000214

(B) 0.0278

(C) 0.00001

(D) 0.1667

(E) 0.1

ANSWER KEY

1. (D)
2. (A)
3. (C)
4. (C)
5. (C)
6. (E)
7. (B)
8. (C)
9. (D)
10. (A)

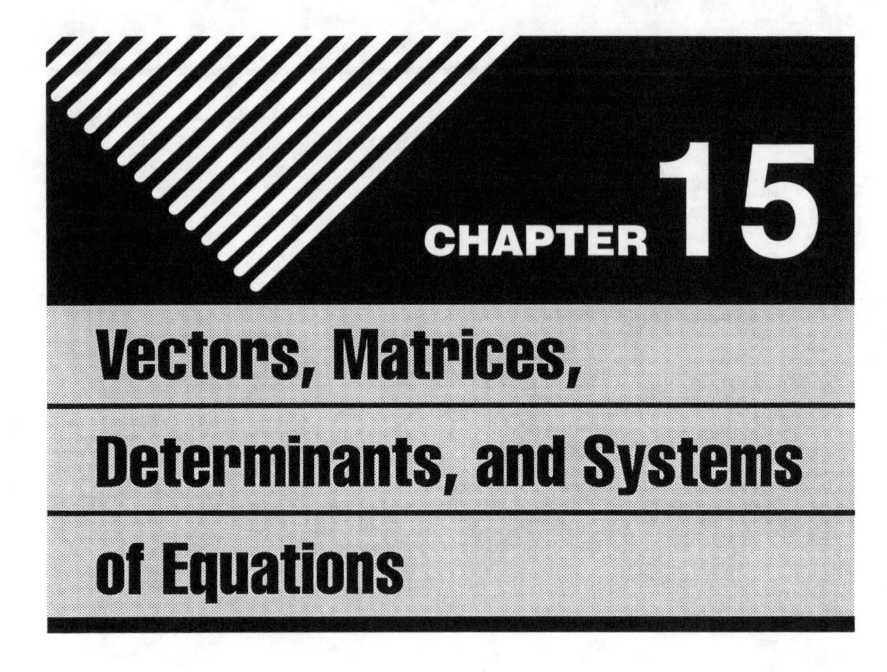

15.1 Vectors

A vector is a quantity having both magnitude and direction, such as displacement, velocity, force, or acceleration.

A scalar is a quantity having magnitude but no direction, e.g., mass, length, time, temperature, or any real number.

A vector in one-, two-, or three-dimensional space can be represented by an arrow; the length of the arrow indicates the magnitude of the vector while its direction indicates the direction of the vector.

Let a vector extend from a point P to a point Q. Then the vector is denoted by $\overrightarrow{PQ}$, and the magnitude is denoted by $|\overrightarrow{PQ}|$.

Two vectors are said to be equal if they have the same magnitude and direction.

If $V(X_0, Y_0, Z_0)$ represents a point in three-dimensional space, and O represents the origin (0, 0, 0), then $\overrightarrow{OV}$ is called the position vector of the point $V(X_0, Y_0, Z_0)$. This vector can be denoted by $\vec{P} = <X_0, Y_0, Z_0>$ where X_0, Y_0, and Z_0 are called the components of $\vec{P}$. The magnitude of $\vec{P}$ is given by

$$|\vec{P}| = \sqrt{X_0^2 + Y_0^2 + Z_0^2}.$$

If $A(X_1, Y_1, Z_1)$ and $B(X_2, Y_2, Z_2)$ are two points in three-dimensional space (three-space), then the vector AB is given by

$$\vec{AB} = <X_2 - X_1, Y_2 - Y_1, Z_2 - Z_1>.$$

If $\vec{p}$, q, and $\vec{r}$ are any vectors in three-space, and a, b, and z are scalars, $z = 0$; let $\vec{n}$ be the null vector, i.e., $|\vec{n}| = 0$ then:

A) $\vec{p} + \vec{q} = \vec{q} + \vec{p}$

B) $\vec{p} - \vec{q} = \vec{p} + (-\vec{q}) = (-\vec{q}) + \vec{p}$

C) $\vec{p} + (\vec{q} + \vec{r}) = (\vec{p} + \vec{q}) + \vec{r}$

D) $\vec{p} + (-\vec{p}) = \vec{n}$

E) $\vec{p} + \vec{n} = \vec{p}$

F) $a(\vec{p} + \vec{q}) = a\vec{p} + a\vec{q} = \vec{p}a + \vec{q}a = (\vec{p} + \vec{q})\, a$

G) $(a + b)\vec{p} = a\vec{p} + b\vec{p} = \vec{p}a + \vec{p}b = \vec{p}(a + b)$

H) $a \times b\vec{p} = a(b\vec{p}) = b(a\vec{p}) = ba \times \vec{p} = \vec{p}(ab)$

I) $z \times \vec{p} = n$

$-\vec{p}$ has the same magnitude as $\vec{p}$ but is opposite in direction.

Algebraically, the sum of two vectors is found by adding the corresponding coordinates of the vectors.

Similarly, the difference of two vectors is found by subtracting corresponding coordinates.

Example: $\vec{a} = <2, 1>$, $\vec{b} = <-3, 5>$

so $\vec{a} + \vec{b} = <2 + (-3), 1 + 5> = <-1, 6>$.

The sum and difference of two vectors can also be found graphically, by the triangle and parallelogram laws of vector addition. Figure 15.1 shown below is self-explanatory.

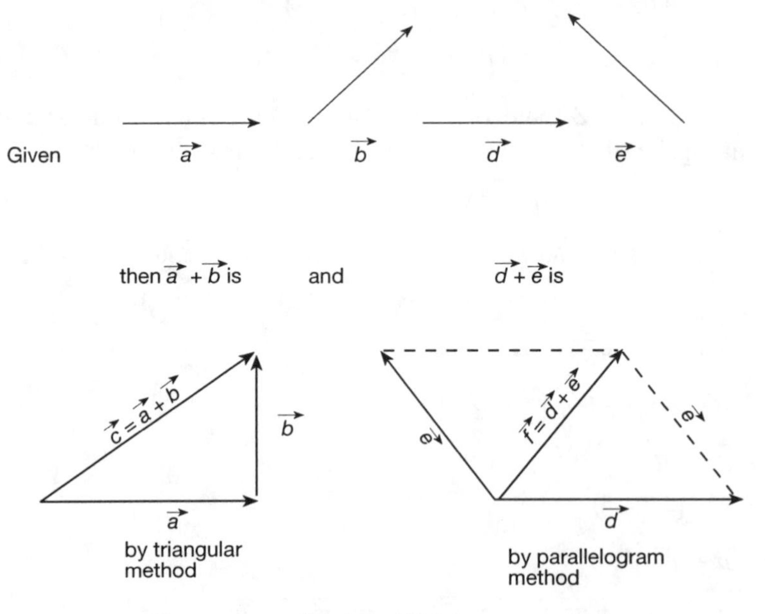

Figure 15.1

Finally, in a rectangular coordinate system in three dimensions, a vector $\overrightarrow{P} = <x_1, y_1, z_1>$ may also be represented as

$$\overrightarrow{p} = x_1 i + y_1 j + z_1 k$$

where $\overrightarrow{i} = <1, 0, 0>$, $\overrightarrow{j} = <0, 1, 0>$ and $\overrightarrow{k} = <0, 0, 1>$. They are called unit vectors.

Problem Solving Examples:

Which of the following vectors are equal to $\overrightarrow{MN}$ in Figure 15.2 if $M = (2, 1)$ and $N = (3, -4)$?

(A) $\overrightarrow{AB}$, where $A = (1, -1)$ and $B = (2, 3)$

(B) $\overrightarrow{CD}$, where $C = (-4, 5)$ and $D = (-3, 10)$

(C) $\overrightarrow{EF}$, where $E = (3, -2)$ and $F = (4, -7)$

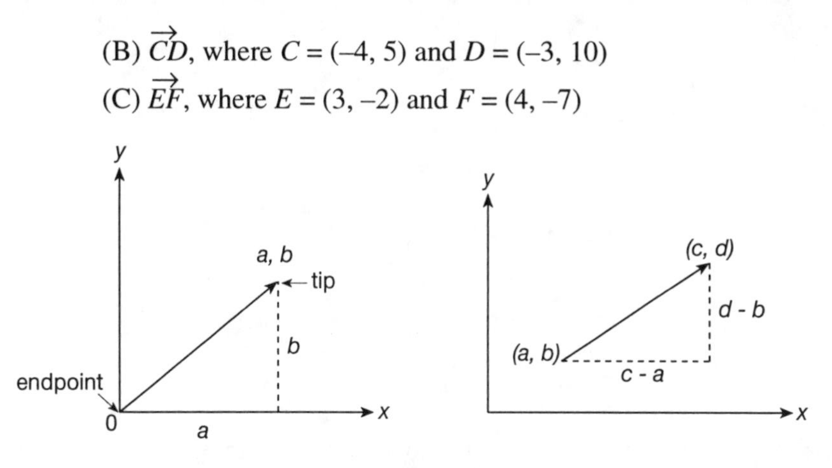

Figure 15.2

With each ordered pair in the plane, there can be associated a vector from the origin to that point.

The vector is determined by subtracting the coordinates of the endpoint from the corresponding coordinates of the tip. As for $\overrightarrow{MN}$, the tip is the point corresponding to the second letter of the alphabetical notation, N, while the endpoint is the point corresponding to the first, M. In this problem the vectors are of a general nature wherein their endpoints do not lie at the origin.

We first find the ordered pair which represents $\overrightarrow{MN}$.

$$\overrightarrow{MN} = (3 - 2, -4 - 1) = (1, -5)$$

Now, we find the ordered pair representing each vector.

(A) $\overrightarrow{AB} = (2 - 1, 3 - (-1)) = (1, 4)$

(B) $\overrightarrow{CD} = ((-3) - (-4), 10 - 5) = (1, 5)$

(C) $\overrightarrow{EF} = (4 - 3, (-7) - (-2)) = (1, -5)$

Only $\overrightarrow{EF}$ and $\overrightarrow{MN}$ are equal.

A force of 315 lbs. is acting at an angle of 67° with the horizontal. What are its horizontal and vertical components?

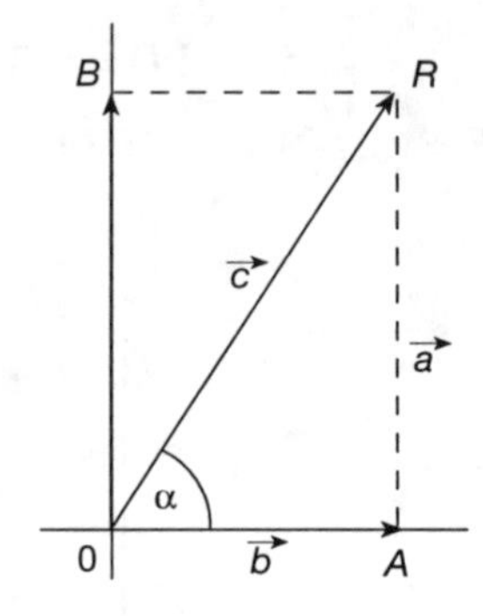

Figure 15.3

Construct the figure as shown in Figure 15.3.

OR = vector force = c.

b = OA = horizontal component.

a = OB = vertical component.

In ΔOAR: $c = 315$; $\alpha = 67°$		
$\frac{a}{c} = \sin\alpha$, or $a = c \sin \alpha$	log c= 2.49831 log sin α= 9.96403 – 10 ――――― log α= 2.46234	a = 289.96 lbs
$\frac{b}{c} = \cos\alpha$, or $b = c \cos \alpha$	log c = 2.49831 log cos α= 9.59188 – 10 ――――― log b = 2.09019	b = 123.08 lbs

15.2 Matrices

A matrix is a rectangular array of numbers, or variables.

Examples:

$$\begin{pmatrix} 2 & 2 & 1 \\ 1 & 0 & 4 \end{pmatrix}, \begin{pmatrix} a_{11} & a_{12} & a_{13} \\ a_{21} & a_{22} & a_{23} \\ a_{31} & a_{32} & a_{33} \\ a_{41} & a_{42} & a_{43} \end{pmatrix}, \begin{pmatrix} 8 & 4 & \frac{5}{3} & 7 \\ 9 & 0 & 1 & -1 \\ 2 & 3 & 4 & 5 \end{pmatrix}.$$

Each number in a given matrix is called an element of the matrix. An element may be denoted by a_{ij}, where i indicates the row containing the element and j indicates the column containing the element.

A matrix of m rows and n columns is said to be a matrix of order m by n $(m \times n)$. An $n \times n$ matrix is called a square matrix. The principal diagonal of a square matrix is the diagonal containing the elements from the upper left-hand corner to the lower right-hand corner.

The identity matrix I of order $n \times n$ is a matrix whose elements are all zeros except elements on the principal diagonal which are ones.

Columns →

$$I_{3\times 3} = \downarrow \text{Rows} \begin{pmatrix} 1 & 0 & 0 \\ 0 & 1 & 0 \\ 0 & 0 & 1 \end{pmatrix}$$

A square matrix is said to be diagonal if entries off its principal diagonal are zero and entries on the principal diagonal are any real numbers.

A matrix is in triangular form if all the elements above or below the principal diagonal are zeros.

Given a system of equations of the form,

$$\begin{array}{ccccccccccc} a_{11}x_1 & + & a_{12}x_2 & + & a_{13}x_3 & + & \dots & + & a_{1n}x_n & = & c_1 \\ a_{21}x_1 & + & a_{22}x_2 & + & a_{23}x_3 & + & \dots & + & a_{2n}x_n & = & c_2 \\ \vdots & & \vdots & & \vdots & & & & \vdots & & \vdots \\ a_{m1}x_1 & + & a_{m2}x_2 & + & a_{m3}x_3 & + & \dots & + & a_{mn}x_n & = & c_m \end{array}$$

the augmented matrix of this system of equations is

$$\left(\begin{array}{ccccc|c} a_{11} & a_{12} & a_{13} & \cdots & a_{1n} & c_1 \\ a_{21} & a_{22} & a_{23} & \cdots & a_{2n} & c_2 \\ \vdots & & & & \vdots & \vdots \\ a_{m1} & a_{m2} & a_{m3} & \cdots & a_{mn} & c_m \end{array}\right)$$

and the coefficient matrix of this system is

$$\begin{pmatrix} a_{11} & a_{12} & a_{13} & \cdots & a_{1n} \\ a_{21} & a_{22} & a_{23} & \cdots & a_{2n} \\ \vdots & \vdots & \vdots & & \vdots \\ a_{m1} & a_{m2} & a_{m3} & \cdots & a_{mn} \end{pmatrix}.$$

Any of the following operations on a matrix is said to be an elementary row operation.

A) Interchange of two rows of a matrix.

B) Multiplication of each element of a row by the same non-zero constant.

C) Addition of the elements of a row multiplied by a non-zero constant to the corresponding elements of another row.

Two matrices are equal if and only if they are of the same order and have the same entries in each position.

To add or subtract matrices is to add or subtract the corresponding entries of the matrices.

Example:

$$\begin{pmatrix} 9 & 1 & 2 \\ 0 & 1 & 0 \end{pmatrix} + \begin{pmatrix} 0 & 0 & 0 \\ 0 & 1 & 1 \end{pmatrix}$$

$$= \begin{pmatrix} 9+0 & 1+0 & 2+0 \\ 0+0 & 1+1 & 0+1 \end{pmatrix} = \begin{pmatrix} 9 & 1 & 2 \\ 0 & 2 & 1 \end{pmatrix}$$

Product of two matrices. The product C of two matrices, $A = [a_{ij}]_{m\times n}$ and $B = [b_{ij}]_{n\times q}$, which are conformable for multiplication in the order AB is defined by

$$A \times B = C = [c_{ij}]_{m\times q}$$

where

$$c_{ij} = a_{i1}b_{1j} + a_{i2}b_{2j} + \ldots + a_{in}b_{nj} = \sum_{k=1}^{n} a_{ik}b_{kj}.$$

In other words, we say that the (i, j) element c_{ij} of the product matrix $C = A \times B$ is the sum of the products of the elements in the ith row of A and the corresponding elements in the jth column of B.

Example:

$$A = \begin{bmatrix} 3 & -5 \\ 7 & 0 \end{bmatrix} \qquad B = \begin{bmatrix} 2 & 4 \\ -8 & 9 \end{bmatrix}$$

Then

$$AB = \begin{bmatrix} 3 & -5 \\ 7 & 0 \end{bmatrix}\begin{bmatrix} 2 & 4 \\ -8 & 9 \end{bmatrix} = \begin{bmatrix} 3(2)+(-5)(-8) & 3(4)+(-5)9 \\ 7(2)+(0)(-8) & 7(4)+(0)9 \end{bmatrix}$$

$$= \begin{bmatrix} 46 & -33 \\ 14 & 28 \end{bmatrix}$$

$$BA = \begin{bmatrix} 2 & 4 \\ -8 & 9 \end{bmatrix}\begin{bmatrix} 3 & -5 \\ 7 & 0 \end{bmatrix} = \begin{bmatrix} 2(3)+4(7) & 2(-5)+4(0) \\ -8(3)+9(7) & (-8)(-5)+9(0) \end{bmatrix}$$

$$= \begin{bmatrix} 34 & -10 \\ 39 & 40 \end{bmatrix}.$$

In this case $AB \neq BA$. Hence, matrix multiplication is not communicative.

Transpose of a matrix. The matrix A^T of order $n \times m$, obtained by interchanging the rows and columns of an $m \times n$ matrix A, is defined as the transpose of A. In symbolic terms, if $A = [a_{ij}]_{m\times n}$, then $A^T = {}_{ji}[a_{ij}]_{n\times m}$.

Example:

If $$A = \begin{bmatrix} a & b \\ c & d \end{bmatrix}$$

then

$$A^T = \begin{bmatrix} a & c \\ b & d \end{bmatrix}.$$

If A, B, and C are any three $m \times n$ matrices, and a, b, and c are real constants, let Z denote the zero matrix then

A) $(A \pm B) \pm C = A \pm (B \pm C)$,

B) $A + (-A) = Z$, where Z is the zero matrix with all entries equal to zero,

C) $A + Z = A$,

D) $(a + b)A = aA + bA$,

E) $a(A + B) = aA + aB$,

F) $(ab)A = b(aA) = a(bA)$, and

G) $I_{m\times m} A_{m\times m} = A$, where $I_{m\times m}$ is the identity matrix.

Problem Solving Example:

If $A = \begin{bmatrix} a_1 & a_2 & a_3 \\ b_1 & b_2 & b_3 \end{bmatrix}$ and $B = \begin{bmatrix} x \\ y \\ z \end{bmatrix}$ find $A \times B$.

A matrix is a set of numbers in a rectangular arrangement. The numbers which make up a matrix are its elements. Matrix A has two rows and three columns; it is called a 2×3 matrix, the number of rows being written first. Matrix B has three rows and one column; it is called a 3×1 matrix. The product of an $m \times n$ matrix A by an $n \times p$ matrix B is an $m \times p$ matrix whose element in the ith row and jth column is the single element of the product of the ith vector of A by the jth column vector of B. An ith row vector is a $1 \times n$ matrix of the form

$$[a_{i1}\ a_{i2}\ \dots\ a_{in}].$$

A jth column vector is $n \times 1$ like $\begin{bmatrix} b_{1j} \\ b_{2j} \\ \vdots \\ b_{nj} \end{bmatrix}$.

$$i \times j = [a_{i1}\ \ a_{i2}\ \ \dots\ \ a_{in}] \begin{bmatrix} b_{1j} \\ b_{2j} \\ \vdots \\ b_{nj} \end{bmatrix} = [a_{i1}\ \ b_{1j}\ \ \dots\ \ a_{in}\ \ b_{nj}].$$

Since A is 2×3, and B is 3×1, $A \times B$ is 2×1.

$$A \times B = \begin{bmatrix} a_1x & + & a_2y & + & a_3z \\ b_1x & + & b_2y & + & b_3z \end{bmatrix}$$

The first row is $a_1x + a_2y + a_3z$. The second row is $b_1x + b_2y + b_3z$. The one column is

$$a_1x + a_2y + a_3z$$
$$b_1x + b_2y + b_3z\ .$$

If the number of columns in A is not equal to the number of rows in B, the product $A \times B$ is not defined. Furthermore, if A and B are square matrices (matrices which have the same number of rows as columns), $A \times B$ is usually not equal to $B \times A$.

15.3 Determinants

For a 2×2 matrix, the determinant is given by:

$$\begin{vmatrix} a_1 & b_1 \\ a_2 & b_2 \end{vmatrix} = a_1b_2 - b_1a_2.$$

For a 3×3 matrix, the determinant is given by:

$$\begin{vmatrix} a_1 & b_1 & c_1 \\ a_2 & b_2 & c_2 \\ a_3 & b_3 & c_3 \end{vmatrix} = a_1b_2c_3 + b_1c_2a_3 + c_1a_2b_3 - c_1b_2a_3 - a_1c_2b_3 - b_1a_2c_3 \ .$$

Another way to find the determinant of a 3×3 matrix is as follows:

A) Rewrite the first two columns on the right of the matrix.

B) Compute the products of the three diagonals running from left to right; prefix each term by a positive sign.

C) Compute the products of the three diagonals running from right to left; prefix each term by a negative sign.

D) The sum of the six products is the value of the determinant.

Figure 15.4 may be helpful for remembering this method.

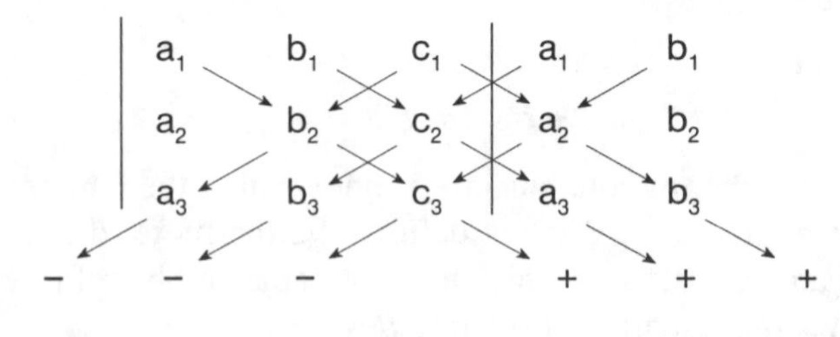

Figure 15.4

A third method for finding the determinant of a 3 × 3 matrix

$$\begin{pmatrix} a_1 & a_2 & a_3 \\ b_1 & b_2 & b_3 \\ c_1 & c_2 & c_3 \end{pmatrix}$$ is the following:

$$\det \begin{pmatrix} a_1 & a_2 & a_3 \\ b_1 & b_2 & b_3 \\ c_1 & c_2 & c_3 \end{pmatrix} = a_1 \begin{vmatrix} b_2 & b_3 \\ c_2 & c_3 \end{vmatrix} - a_2 \begin{vmatrix} b_1 & b_3 \\ c_1 & c_3 \end{vmatrix} + a_3 \begin{vmatrix} b_1 & b_2 \\ c_1 & c_2 \end{vmatrix}.$$

Note that we start with a positive sign, and alternate signs after.

15.3.1 Properties of Determinants

A) If matrix A is obtained from matrix B by interchanging any two rows or columns of B, then $\det A = -\det B$.

B) A determinant does not change its value if a linear combination of other rows (columns) is added to any given row (column).

C) If A is obtained from B by multiplying one row of B by a non-zero constant a, then

$$\det A = a \det B.$$

D) If two rows or columns of a matrix A are identical, then

$$\det A = 0.$$

E) If I is the identity matrix, then

$$\det I = 1.$$

F) If an entire row (or a column) of a matrix consists of only zeroes, the determinant of this matrix is zero.

G) If a square matrix is in triangular form, then the determinant of this matrix is the product of all the numbers on the principal diagonal.

Problem Solving Examples:

Find the value of the determinant $\begin{vmatrix} 1 & 2 \\ 2 & 3 \end{vmatrix}$.

The value of a 2×2 determinant can be found by the following equation:

$$\begin{vmatrix} a & c \\ b & d \end{vmatrix} = ad - bc .$$

Hence, the value of $\begin{vmatrix} 1 & 2 \\ 2 & 3 \end{vmatrix}$ is:

$$\begin{vmatrix} 1 & 2 \\ 2 & 3 \end{vmatrix} = (1)(3) - (2)(2) = 3 - 4 = -1.$$

Evaluate the determinant $\begin{vmatrix} 3 & 5 \\ -2 & 3 \end{vmatrix}$.

The determinant of any 2×2 matrix is

$$\begin{vmatrix} a & b \\ c & d \end{vmatrix} = ad - bc.$$

Apply this rule to the given 2×2 matrix.

$$\begin{vmatrix} 3 & 5 \\ -2 & 3 \end{vmatrix} = (3)(3) - (-2)(5) = 9 + 10 = 19$$

15.4 Systems of Equations

Consider a system of n linear equations in n unknowns:

$$a_{11}x_1 + a_{12}x_2 + \ldots + a_{1n}x_n = b_1$$
$$a_{21}x_1 + a_{22}x_2 + \ldots + a_{2n}x_n = b_2$$
$$\ldots\ldots\ldots\ldots\ldots\ldots\ldots\ldots$$
$$a_{n1}x_1 + a_{n2}x_2 + \ldots + a_{nn}x_n = b_n$$

Transforming the above equations into matrix notation, we get

$$\begin{bmatrix} a_{11} & a_{12} & \cdots & a_{1n} \\ a_{21} & a_{22} & \cdots & a_{2n} \\ \vdots & & & \vdots \\ a_{n1} & a_{n2} & & a_{nn} \end{bmatrix} \begin{bmatrix} x_1 \\ x_2 \\ \vdots \\ x_n \end{bmatrix} = \begin{bmatrix} b_1 \\ b_2 \\ \vdots \\ b_n \end{bmatrix}$$

or, $AX = B$.

Let A be an $n \times n$ matrix over the field F such that $\det A \neq 0$. If $b_1, b_2, \ldots, b_n$ are any scalars in F, the unique solution of the system of equations $AX = B$ is given by:

$$x_i = \frac{\det A_i}{\det A} \qquad i = 1, 2, \ldots, n.$$

where A_i is the $n \times n$ matrix obtained from A by replacing the ith column of A by the column vector.

$$\begin{bmatrix} b_1 \\ b_2 \\ \vdots \\ b_n \end{bmatrix}$$

The above theorem is known as “Cramer’s Rule” for solving systems of linear equations. Cramer’s Rule applies only to systems of n linear equations in n unknowns with non-zero determinants.

Example: Solve the given system of equations, by the method of determinants

$$3x - 5y = 4$$

$$7x + 4y = 25$$

Solution: The equations, as given, are in standard form for applying Cramer's Rule. Therefore,

$$x = \frac{\begin{vmatrix} 4 & -5 \\ 25 & 4 \end{vmatrix}}{\begin{vmatrix} 3 & -5 \\ 7 & 4 \end{vmatrix}} = \frac{4 \times 4 - 25(-5)}{3 \times 4 - 7(-5)} = \frac{16 + 125}{12 + 35} = \frac{141}{47} = 3,$$

$$\text{and } y = \frac{\begin{vmatrix} 3 & 4 \\ 7 & 25 \end{vmatrix}}{\begin{vmatrix} 3 & -5 \\ 7 & 4 \end{vmatrix}} = \frac{3 \times 25 - 7 \times 4}{47} = \frac{75 - 28}{47} = \frac{47}{47} = 1.$$

This process always yields a unique solution unless the denominator determinant is equal to zero.

A system of linear equations is said to be in echelon (or triangular) form if the coefficient matrix of the system is in triangular form.

Example:

$$\begin{aligned} 2x_1 + 4x_2 + x_3 &= 0 \\ x_2 + x_3 &= 5 \\ x_3 &= 4 \end{aligned}$$

A system of k linear equations in k unknowns is determinative if and only if the determinant of the coefficient matrix of this system is not zero.

If the system is determinative, then it is consistent, i.e., contains no contradictions like:

$$x + y = 0$$

$$x + y = 1.$$

Problem Solving Examples:

Solve
$$x + y = 3$$
$$2x + 3y = 1$$

A The values for x and y can be determined by use of Cramer's Rule and determinants. The value for x is the quotient of two determinants. The determinant in the denominator consists of vertical columns in which the numbers are the coefficients of the variables. The determinant in the numerator is the same as the determinant in the denominator, except that the first vertical column is replaced by the constant terms. Note that the first vertical column in the two given equations corresponds to the x term.

$$\begin{array}{ccccc} x & + & y & = & 3 \\ 2x & + & 3y & = & 1 \\ \uparrow & & \uparrow & & \nwarrow \end{array}$$

1st	2nd	3rd
vertical	vertical	vertical
column	column	column

The third vertical column consists of the constant terms. Hence,

$$x = \frac{\begin{vmatrix} 3 & 1 \\ 1 & 3 \end{vmatrix}}{\begin{vmatrix} 1 & 1 \\ 2 & 3 \end{vmatrix}} = \frac{(3)(3)-(1)(1)}{(1)(3)-(2)(1)} = \frac{9-1}{3-2} = \frac{8}{1} = 8.$$

The value for y is also the quotient of two determinants. The determinant in the denominator is the same as the determinant in the denominator used for finding x. The determinant in the numerator is the same as the determinant in the denominator, except that the second vertical column is replaced by the constant terms. Note that the second vertical column in the two given equations corresponds to the y term. (See the illustration.) Hence,

$$y = \frac{\begin{vmatrix} 1 & 3 \\ 2 & 1 \end{vmatrix}}{\begin{vmatrix} 1 & 1 \\ 2 & 3 \end{vmatrix}} = \frac{(1)(1)-(2)(3)}{(1)(3)-(2)(1)} = \frac{1-6}{3-2} = \frac{-5}{1} = -5.$$

Solve the equations $2x + 4y = 11$ and $-5x + 3y = 5$ graphically and by means of determinants.

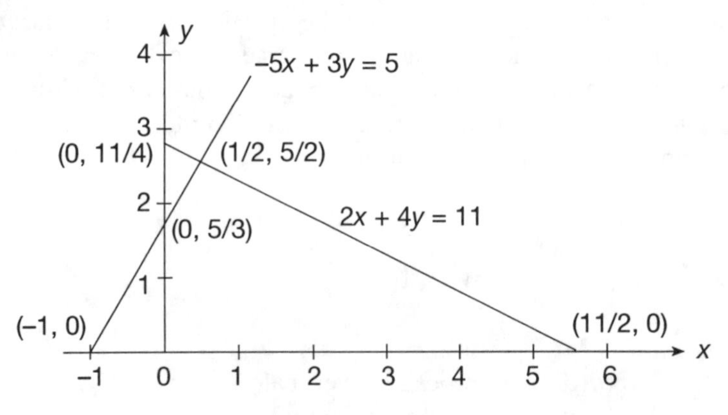

Figure 15.5

A) To solve a set of linear equations graphically (Figure 15.5), we find their point of intersection (which satisfies both equations simultaneously). Draw both lines by determining their y- and x-intercepts by setting $x = 0$ and $y = 0$ respectively. See the following tables. Note: We solve for y before finding the y-intercept and solve for x to find the x-intercept.

x	$2x + 4y = 11$	y
0	$4y = 0$	$\frac{11}{4}$
$\frac{11}{2}$	$2x = 11$	0

x	$-5x + 3y = 5$	y
0	$3y = 5$	$\frac{5}{3}$
–1	$-5x = 5$	0

Now each line can be plotted from two points. We see from the graph that the point of intersection seems to be approximately $\left(\frac{1}{2}, \frac{5}{2}\right)$.

B) To solve a system of two linear equations in two unknowns by determinants, we set up the following solutions in determinant form, derived from the linear equations in standard form (see Figure 15.6).

$$a_1x + b y = c_1$$

$$a_2x + b_2y = c_2$$

$$x = \frac{\begin{vmatrix} c_1 & b_1 \\ c_2 & b_2 \end{vmatrix}}{\begin{vmatrix} a_1 & b_1 \\ a_2 & b_2 \end{vmatrix}} \qquad y = \frac{\begin{vmatrix} a_1 & c_1 \\ a_2 & c_2 \end{vmatrix}}{\begin{vmatrix} a_1 & b_1 \\ a_2 & b_2 \end{vmatrix}}$$

Figure 15.6

The denominator for both variables is formed by writing the coefficients of x and y in the linear equations. The numerators are formed from the denominator by replacing the column of coefficients of that unknown by the column of constants.

In this case the linear equations in standard form are:

$$2x + 4y = 11$$

$$-5x + 3y = 5.$$

Then, the solution by determinants is

$$x = \frac{\begin{vmatrix} 11 & 4 \\ 5 & 3 \end{vmatrix}}{\begin{vmatrix} 2 & 4 \\ -5 & 3 \end{vmatrix}}.$$

The value of a 2×2 determinant is defined to be:

$$\begin{vmatrix} a & b \\ c & d \end{vmatrix} = ad - bc.$$

Therefore,

$$x = \frac{(11)(3)-(4)(5)}{(2)(3)-(4)(-5)} = \frac{33-20}{6-(-20)} = \frac{13}{26} = \frac{1}{2};$$

$$y = \frac{\begin{vmatrix} 2 & 11 \\ -5 & 5 \end{vmatrix}}{26} = \frac{(2)(5)-(-5)(11)}{26} = \frac{10-(-55)}{26} = \frac{65}{26} = \frac{5}{2}.$$

This agrees with the graphical solution and confirms that the point of intersection is $\left(\frac{1}{2}, \frac{5}{2}\right)$.

CHAPTER 16

Mathematical Induction and the Binomial Theorem

16.1 Mathematical Induction

Mathematical induction is a method of proof. The steps are:

A) Verification of the proposed formula or theorem for the smallest value of n.

B) Assume that the theorem is true for $n = k$ then prove that it is true for $n = k + 1$.

C) Conclude that the proposed theorem holds true for all values of n.

Example: Prove by mathematical induction that

$1 + 7 + 13 + \ldots + (6n - 5) = n(3n - 2)$.

Solution:

(A) The proposed formula is true for $n = 1$ since $1 = 1(3 - 2)$.

(B) Assume the formula to be true for $n = k$, a positive integer; that is, assume

(a) $1 + 7 + 13 + \ldots + (6k - 5) = k(3k - 2)$.

Under this assumption we wish to show that

(b) $1 + 7 + 13 + \ldots + (6k - 5) + [6(k + 1) - 5] = (k + 1)[3(k + 1) - 2]$, which is equivalent to when $(6k + 1)$ is added to both sides of a), we have on the right $k(3k - 2) + (6k + 1) = 3k^2 + 4k + 1 = (k + 1)(3k + 1)$; hence, if the formula is true for $n = k$, it is true for $n = k + 1$.

(C) Since the formula is true for $n = k = 1$ (Step A), it is true for $n = k + 1 = 2$; being true for $n = k = 2$ it is true for $n = k + 1 = 3$; and so on, for every positive integral value of n.

Definition: The expression $n!$ is read "n factorial" and is the product of all consecutive integers from n down to 1.

Examples: $3! = 3 \times 2 \times 1$

$7! = 7 \times 6 \times 5 \times 4 \times 3 \times 2 \times 1$

Note: $0! = 1$ by definition.

Problem Solving Examples:

Prove by mathematical induction

$$1^2 + 2^2 + 3^2 + \ldots + n^2 = \frac{1}{6}n(n + 1)(2n + 1).$$

(A) Using mathematical induction, first verify that the proposed formula is true for the smallest value of n. It is desirable, but not necessary, to verify it for several values of n.

(B) The proof that if the proposed formula or theorem is true for $n = k$, some positive integer, it is true also for $n = k + 1$. That is, if the proposition is true for any particular value of n, it must be true for the next larger value of n.

(C) A conclusion that the proposed formula holds true for all values of n.

Step 1: Verify:

For $n = 1$:
$$1^2 = \frac{1}{6}(1)(1+1)[2(1)+1] = \frac{1}{6}(1)(2)(3)$$
$$= \frac{1}{6}(6) = 1$$
$$1 = 1 \checkmark$$

For $n = 2$:
$$1^2 + 2^2 = \frac{1}{6}(2)(2+1)[2(2)+1] = \frac{1}{6}(2)(3)(5) = \frac{1}{6}(6)(5)$$
$$1 + 4 = (1)(5)$$
$$5 = 5 \checkmark$$

For $n = 3$:
$$1^2 + 2^2 + 3^2 = \frac{1}{6}(3)(3+1)[2(3)+1]$$
$$1 + 4 + 9 = \frac{1}{6}(3)(4)(7) = \frac{1}{6}(12)(7) = 14$$
$$14 = 14 \checkmark$$

Step 2. Let k represent any particular value of n. For $n = k$, the formula becomes

$$1^2 + 2^2 + 3^2 + \ldots + k^2 = \frac{1}{6}k(k+1)(2k+1). \qquad (1)$$

For $n = k + 1$, the formula is

$$1^2 + 2^2 + 3^2 + \ldots + k^2 + (k+1)^2 = \frac{1}{6}(k+1)[(k+1)+1][2(k+1)+1]$$
$$= \frac{1}{6}(k+1)(k+2)(2k+3) \qquad (2)$$

We must show that if the formula is true for $n = k$, then it must be true for $n = k + 1$. In other words, we must show that Equation (2) follows from Equation (1). The left side of Equation (1) can be converted into the left side of Equation (2) by merely adding $(k + 1)^2$. All

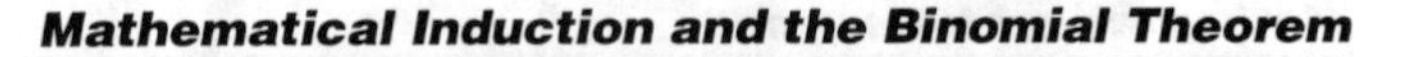

that remains to be demonstrated is that when $(k + 1)^2$ is added to the right side of Equation (1), the result is the right side of Equation (2).

$$1^2 + 2^2 + \ldots + k^2 + (k + 1)^2 = \frac{1}{6}k(k + 1)(2k + 1) + (k + 1)^2$$

Factor out $(k + 1)$:

$$\begin{aligned} 1^2 + 2^2 + 3^2 + \ldots + k^2 + (k + 1)^2 &= (k + 1)\left[\frac{1}{6}k(2k+1)+(k+1)\right] \\ &= (k+1)\left[\frac{k(2k+1)}{6}+\frac{(k+1)6}{6}\right] \\ &= (k+1)\frac{2k^2+k+6k+6}{6} \\ &= \frac{(k+1)(2k^2+7k+6)}{6} \\ &= \frac{1}{6}(k+1)(k+2)(2k+3), \end{aligned}$$

since

$$2k^2 + 7k + 6 = (k + 2)(2k + 3).$$

Thus, we have shown that if we add $(k + 1)^2$ to both sides of the equation for $n = k$, then we obtain the equation or formula for $n = k + 1$. We have thus established that if Equation (1) is true, then Equation (2) must be true; that is, if the formula is true for $n = k$, then it must be true for $n = k + 1$. In other words, we have proved that if the proposition is true for a certain positive integer k, then it is also true for the next greater integer $k + 1$.

Step 3. The proposition is true for $n = 1, 2, 3$ (Step 1). Since it is true for $n = 3$, it is true for $n = 4$ (Step 2, where $k = 3$ and $k + 1 = 4$). Since it is true for $n = 4$, it is true for $n = 5$, and so on, for all positive integers n.

Prove by mathematical induction that the sum of n terms of an arithmetic progression a, $a + d$, $a + 2d$, … is $\frac{n}{2}[2a+(n-1)d]$, that is

$$a + (a + d) + (a + 2d) + \ldots + [a+(n-1)d] = \frac{n}{2}[2a+(n-1)d].$$

Step 1. The formula holds for $n = 1$, since

$$a = \frac{1}{2}[2a+(1-1)d] = a.$$

Step 2. Assume that the formula holds for $n = k$. Then

$$a + (a + d) + (a + 2d) + \ldots + [a + (k - 1)d] = \frac{k}{2}[2a + (k - 1)d].$$

Add the $(k + 1)$th term, which is $(a + kd)$, to both sides of the latter equation. Then

$$a+(a+d)(a+2d)+\ldots+[a+(k-1)d]+(a+kd)=\frac{k}{2}[2a+(k-1)d]+(a+kd).$$

The right-hand side of this equation $= ka + \frac{k^2d}{2} - \frac{kd}{2} + a + kd$

$$= \frac{k^2d+kd+2ka+2a}{2}$$

$$= \frac{kd(k+1)+2a(k+1)}{2}$$

$$= \frac{k+1}{2}(2a+kd)$$

which is the value of $\frac{n}{2}[2a+(n-1)d]$ when n is replaced by $(k + 1)$.

Hence, if the formula is true for $n = k$, we have proved it to be true for $n = k + 1$. But the formula holds for $n = 1$; hence, it holds for $n = 1 + 1 = 2$. Then, since it holds for $n = 2$, it holds for $n = 2 + 1 = 3$, and so on. Thus, the formula is true for all positive integral values of n.

16.2 Binomial Theorem

The following equation is the binomial theorem or binomial expansion:

$$(x+y)^n = \sum_{k=0}^{n} \binom{n}{k} x^{n-k} y^k$$

$$= \binom{n}{0} x^n + \binom{n}{1} x^{n-1} y + \binom{n}{2} x^{n-2} y^2 +$$

$$\ldots + \binom{n}{k} x^{n-k} y^k + \binom{n}{n} y^n$$

Example: Find the expansion of $(a - 2x)^7$.

Solution: Use the binomial formula:

$$(u+v)^n = u^n + nu^{n-1}v + \frac{n(n-1)}{2}u^{n-2}v^2$$

$$+ \frac{n(n-1)(n-2)}{2 \times 3}u^{n-3}v^3 + \ldots + v^n$$

and substitute a for u and $(-2x)$ for v and 7 for n to obtain:

$(a - 2x)^7 = [a + (-2x)]^7$

$$= a^7 + 7a^6(-2x) + \frac{7 \times 6}{2}a^5(-2x)^2 + \frac{7 \times 6 \times 5}{2 \times 3}a^4(-2x)^3$$

$$+ \frac{7 \times 6 \times 5 \times 4}{2 \times 3 \times 4}a^3(-2x)^4 + \frac{7 \times 6 \times 5 \times 4 \times 3}{2 \times 3 \times 4 \times 5}a^2(-2x)^5$$

$$+ \frac{7 \times 6 \times 5 \times 4 \times 3 \times 2}{2 \times 3 \times 4 \times 5 \times 6}a^1(-2x) + \frac{7 \times 6 \times 5 \times 4 \times 3 \times 2 \times 1}{2 \times 3 \times 4 \times 5 \times 6 \times 7}a^0(-2x)^7$$

$$(a - 2x)^7 = a^7 - 14a^6x + 84a^5x^2 - 280a^4x^3 + 560a^3x^4 - 672a^2x^5$$
$$+ 448ax^6 - 128x^7.$$

16.2.1 Pascal's Triangle

The coefficients of $(a + b)^0$, $(a + b)^1$, $(a + b)^2$ …, $(a + b)^n$ can be obtained from Pascal's Triangle (Figure 16.1):

$(a+b)^0$	1
$(a+b)^1$	1 1
$(a+b)^2$	1 2 1
$(a+b)^3$	1 3 3 1
$(a+b)^4$	1 4 6 4 1
$(a+b)^5$	1 5 10 10 5 1
$(a+b)^6$	1 6 15 20 15 6 1
$(a+b)^7$	1 7 21 35 35 21 7 1
$(a+b)^8$	1 8 28 56 70 56 28 8 1
$(a+b)^9$	1 9 36 84 126 126 84 36 9 1

Figure 16.1

where each number in the triangle is the sum of the two numbers above it, or one if it is on the edge.

Problem Solving Examples:

Q Expand $(2z - 3y)^4$.

Use the binomial theorem.

$$(u+v)^n = u^n + nu^{n-1}v + \frac{n(n-1)}{2}v^{n-2}v^2 + \frac{n(n-1)(n-2)}{2x3}n - 2v^3 + \ldots + v^n$$

where $u = 2z$
$v = -3y$
$n = 4.$

Therefore,

$$
\begin{aligned}
(2z - 3y)^4 &= (2z)^4 + 4(2z)^3(-3y) + 6(2z)^2(-3y)^2 + 4(2z)(-3y)^3 + (-3y)^4 \\
&= 16z^4 + 4(8z^3)(-3y) + 6(4z^2)(9y^2) + 4(2z)(-27y^3) + 81y^4 \\
&= 16z^4 - 96z^3y + 216z^2y^2 - 216zy^3 + 81y^4.
\end{aligned}
$$

Find the first five terms of the expansion of $(1 + x)^{-2}$.

The binomial theorem states that:

$$(a+b)^n = \frac{1}{0!}a^n + \frac{n}{1!}a^{n-1}b + \frac{n(n-1)}{2!}a^{n-2}b^2 + \ldots + nab^{n-1} + b^n.$$

This theorem can be used to find the first five terms of the expansion of $(1 + x)^{-2}$. Replacing a by 1 and b by x, the expression $(1 + x)^{-2}$ becomes:

$$
\begin{aligned}
(1+x)^{-2} = &\frac{1}{0!}1^{-2} + \frac{-2}{1!}1^{-3}x + \frac{(-2)(-3)}{2!}1^{-4}x^2 + \frac{(-2)(-3)(-4)}{3!}1^{-5}x^3 \\
&+ \frac{(-2)(-3)(-4)(-5)}{4!}1^{-6}x^4 + \ldots + (-2)1x^{-3} + x^{-2}.
\end{aligned}
$$

Writing only the first five terms of this expansion:

$$
\begin{aligned}
(1+x)^{-2} &= \frac{1}{0!}1^{-2} + \frac{-2}{1!}1^{-3}x + \frac{(-2)(-3)}{2!}1^{-4}x^2 + \frac{(-2)(-3)(-4)}{3!}1^{-5}x^3 \\
&\quad \frac{(-2)(-3)(-4)(-5)}{4!}1^{-6}x^4 + \ldots \\
&= \frac{1}{1}\left(\frac{1}{1^2}\right) - 2x\left(\frac{1}{1^3}\right) + \frac{6x^2}{2\times 1}\left(\frac{1}{1^4}\right) + \frac{(-24)x^3}{3\times 2\times 1}\left(\frac{1}{1^5}\right) \\
&\quad + \frac{120x^4}{4\times 3\times 2\times 1}\left(\frac{1}{1^6}\right) + \ldots
\end{aligned}
$$

$$(1 + x)^{-2} = 1 - 2x + 3x^2 - 4x^3 + 5x^4 + \ldots \qquad (1)$$

Hence, the right side of Equation (1) represents the first five terms of the expansion of $(1 + x)^{-2}$.

Quiz: Vectors, Matrices—Mathematical Induction

1. Find the coefficient of x^2y^3 in the binomial expansion of $(x - 2y)^5$.

   (A) –160
   (B) 80
   (C) –80
   (D) 8
   (E) –8

2. Find the expansion of $(x + y)^6$.

   (A) $x^6 + 3x^4y^2 + 3x^2y^4 + y^6$
   (B) $x^6 + y^6$
   (C) $x^6 + 5x^5y + 10x^4y^2 + 10x^3y^3 + 5x^2y + y^6$
   (D) $x^6 + 6x^5y + 15x^4y^2 + 20x^3y^3 + 15x^2y^4 + 6xy^5 + y^6$
   (E) $x^6 - 6x^5y - 15x^4y^2 - 20x^3y^3 - 15x^2y^4 - 6xy^5 - y^6$

3. If $A = \begin{bmatrix} 3 & -5 \\ 7 & 0 \end{bmatrix}$ and $B = \begin{bmatrix} 2 & 4 \\ -8 & 9 \end{bmatrix}$, find AB.

   (A) $\begin{bmatrix} 46 & -33 \\ 14 & 28 \end{bmatrix}$
   (B) $\begin{bmatrix} 6 & 20 \\ -56 & 0 \end{bmatrix}$
   (C) $\begin{bmatrix} -46 & 33 \\ 14 & 28 \end{bmatrix}$
   (D) $\begin{bmatrix} 34 & -10 \\ 39 & 40 \end{bmatrix}$
   (E) $\begin{bmatrix} 34 & 10 \\ 39 & 40 \end{bmatrix}$

4. Evaluate the determinant: $\begin{bmatrix} 2 & -1 & 2 \\ 3 & 3 & 6 \\ 5 & 0 & -1 \end{bmatrix}$.

(A) 69
(B) 72
(C) –63
(D) –69
(E) –66

5. Evaluate, or expand, the determinant: $\begin{bmatrix} 2 & 3 \\ 3 & -1 \end{bmatrix}$.

(A) 7
(B) –11
(C) 11
(D) –7
(E) None of the above.

6. Obtain the product $B \times A$ if $A = \begin{bmatrix} 3 & -2 \\ 1 & 4 \end{bmatrix}$ and $B = \begin{bmatrix} 5 & 1 \\ -2 & 3 \end{bmatrix}$.

(A) $\begin{bmatrix} 19 & -3 \\ -3 & 13 \end{bmatrix}$

(B) $\begin{bmatrix} 15 & -2 \\ -2 & 12 \end{bmatrix}$

(C) $\begin{bmatrix} 16 & -6 \\ -3 & 16 \end{bmatrix}$

(D) $\begin{bmatrix} 3 & -10 \\ 3 & -8 \end{bmatrix}$

(E) Not given.

7. Find the value of the determinant of $\begin{vmatrix} 67 & 19 & 21 \\ 39 & 13 & 14 \\ 81 & 24 & 26 \end{vmatrix}$.

(A) 83
(B) 43
(C) –83
(D) –63
(E) – 43

8. Use Cramer's rule to solve the system of equations:

$$3x + y - 2z = -3$$
$$2x + 7y + 3z = 9$$
$$4x - 3y - z = 7$$

(A) {–2, 1, 4}
(B) {–2, –1, 4}
(C) {2, 1, 4}
(D) {2, –1, 4}
(E) {–2, –1, – 4}

9. What is the coefficient of the fifth term in the expansion of $(2x - y)^4$?

(A) –1
(B) 1
(C) – 4
(D) 4
(E) 5

10. Find the coefficient of the fourth term in the expansion of $(j - k)^5$.

(A) 10
(B) –10
(C) 5
(D) 4
(E) – 4

ANSWER KEY

1. (C)
2. (D)
3. (A)
4. (D)
5. (B)
6. (C)
7. (E)
8. (D)
9. (B)
10. (B)

CHAPTER 17

Partial Fractions

If the degree of the numerator of a polynomial fraction is less than that of the denominator, then the fraction is said to be proper; otherwise, it is said to be improper.

To decompose a given proper fraction into partial fractions is to resolve the fraction into a sum of simpler proper fractions.

Let $f(x)$ and $g(x)$ be polynomials and the degree of $f(x)$ be less than that of $g(x)$. To decompose $\frac{f(x)}{g(x)}$ into partial fractions is to find

$$\frac{f(x)}{g(x)} = p_1 + p_2 + \ldots + p_r$$

where each p_i has one of the forms

$$\frac{A}{(ux+v)^m} \quad \text{or} \quad \frac{Bx+C}{(ax^2+bx+c)^n}$$

where $b^2 - 4ac < 0$, that is $ax^2 + bx + c$ is irreducible, and m and n are non-negative integers.

The method for decomposing a given rational fraction is given on the following page.

Step 1: If the degree of the numerator is greater than that of the denominator, rearrange the given fraction so that it is expressed as a sum of a polynomial and a proper rational fraction. (This can often be done through long division.)

Step 2: Express the denominator of the proper rational fraction as a product of different factors of the form $(ux + v)^m$ and/or $(ax^2 + bx + c)^n$ where $ax^2 + bx + c$ is irreducible, and m and n are non-negative integers.

Step 3: For each factor of the form $(ux + v)^m$, $m \geq 1$, the decomposition of the proper rational fraction contains a sum of m partial fractions of the form

$$\frac{A_1}{ux+v}+\frac{A_2}{(ux+v)^2}+\ \dots\ +\frac{A_m}{(ux+v)^m}$$

where each A_i is a real constant to be found later.

For each factor of the form $(ax^2 + bx + c)^n$, $n \geq 1$ and $b^2 - 4ac < 0$, the decomposition of the proper rational fraction contains a sum of n partial fractions of the form

$$\frac{A_1x+B_1}{ax^2+bx+c}+\frac{A_2x+B_2}{(ax^2+bx+c)^2}+\ \dots\ +\frac{A_nx+B_n}{(ax^2+bx+c)^n}$$

where each A_i and B_i are real constants to be found in the next step.

Therefore, the proper rational fraction $\frac{f(x)}{g(x)}$ is now in the form

$$\frac{f(x)}{g(x)} = \underbrace{p_1 + p_2 + \dots + p_r}_{\text{partial fractions}}.$$

Step 4: Find the common denominator of $p_1, p_2, \dots, p_r$.

Then express each of the partial fractions obtained in Step 3 as fractions $p_1^*, p_2^*, \dots, p_r^*$ all having a common denominator $q(x)$. Thus, we have,

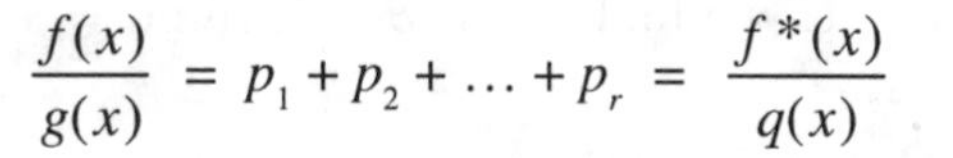

$$\frac{f(x)}{g(x)} = p_1 + p_2 + \ldots + p_r = \frac{f^*(x)}{q(x)}$$

where $g(x) = q(x)$ and $f^*(x)$ is the sum of the numerators of p_1^*, p_2^*, ..., p_r^*. Using the identity $f(x) = f^*(x)$, we can obtain a system of equations which enables us to solve for $A_1, A_2, \ldots, A_m, A_n, B_1, B_2, \ldots, B_m, \ldots$ etc. as illustrated by the following example.

Example: Decompose

$$\frac{3x^2+2x-2}{x^3-1}$$

into partial fractions.

Solution: The denominator can be factored into the product

$$x^3 - 1 = (x-1)(x^2+x+1),$$

and we write:

$$\frac{3x^2+2x-2}{x^3-1} = \frac{A}{x-1} + \frac{Bx+C}{x^2+x+1}$$

$$= \frac{A(x^2+x+1)+(Bx+C)(x-1)}{(x-1)(x^2+x+1)}.$$

Setting the numerators of the above fractions equal, we have:

$$3x^2 + 2x - 2 = A(x^2 + x + 1) + (Bx + C)(x - 1).$$

Now we multiply out and collect like powers of x. We obtain:

$$3x^2 + 2x - 2 = (A + B)x^2 + (A - B + C)x + (A - C).$$

Equating coefficients of like powers of x, we obtain:

$$A + B = 3$$

$$A - B + C = 2$$

$$A - C = -2.$$

Solving for A, B, and C, we find $A = 1$, $B = 2$, and $C = 3$. Therefore, we have:

$$\frac{3x^2+2x-2}{x^3-1} = \frac{1}{x-1} + \frac{2x+3}{x^2+x+1}.$$

Problem Solving Examples:

Q Combine $\dfrac{3x+y}{x^2-y^2} - \dfrac{2y}{x(x-y)} - \dfrac{1}{x+y}$ into a single fraction.

A Fractions that have different denominators must be transformed into fractions with the same denominator before they may be combined. This identical denominator is the least common denominator (LCD), the least common multiple of the denominators of the fractions to be added. In the process of transforming the fractions to a common denominator, we make use of the fact that the numerator and denominator of a fraction may be multiplied by the same non-zero number without changing the value of the fraction. In our case, the denominators are $x^2 - y^2 = (x + y)(x - y)$, $x(x - y)$, and $x + y$. Therefore, the LCD is $x(x + y)(x - y)$, and we proceed as follows:

$$\frac{3x+y}{x^2-y^2} - \frac{2y}{x(x-y)} - \frac{1}{x+y} = \frac{3x+y}{(x+y)(x-y)} - \frac{2y}{x(x-y)} - \frac{1}{x+y}$$

$$= \frac{x(3x+y)}{x(x+y)(x-y)} - \frac{(x+y)2y}{(x+y)(x)(x-y)} - \frac{x(x-y)}{x(x-y)(x+y)}$$

$$= \frac{3x^2+xy}{x(x+y)(x-y)} - \frac{2xy+2y^2}{x(x+y)(x-y)} - \frac{x^2-xy}{x(x+y)(x-y)}$$

$$= \frac{3x^2 + xy - (2xy + 2y^2) - (x^2 - xy)}{x(x+y)(x-y)}$$

$$= \frac{3x^2 + xy - 2xy - 2y^2 - x^2 + xy}{x(x+y)(x-y)}$$

$$= \frac{3x^2 - x^2 + xy + xy - 2xy - 2y^2}{x(x+y)(x-y)}$$

$$= \frac{2x^2 - 2y^2}{x(x+y)(x-y)}$$

$$= \frac{2(x^2 - y^2)}{x(x+y)(x-y)}$$

$$= \frac{2\cancel{(x+y)}\cancel{(x-y)}}{x\cancel{(x+y)}\cancel{(x-y)}}$$

$$= \frac{2}{x}$$

Perform the following addition: $\dfrac{2x}{x^2 - 4} + \dfrac{3}{x^2 - 5x + 6}$.

Factor the denominators into polynomial factors. Hence,

$$x^2 - 4 = (x + 2)(x - 2)$$

$$x^2 - 5x + 6 = (x - 3)(x - 2).$$

Therefore:

$$\frac{2x}{x^2 - 4} + \frac{3}{x^2 - 5x + 6} = \frac{2x}{(x+2)(x-2)} + \frac{3}{(x-3)(x-2)} \qquad (1)$$

Now, find the least common denominator (LCD) of the two fractions on the right side of Equation (1). This is done by writing down all the different factors that appear in the two denominators. The exponent to be used for each factor is the smallest number of times that the factor appears in either denominator. Hence,

$$\text{LCD} = (x+2)^1(x-2)^1(x-3)^1 = (x+2)(x-2)(x-3).$$

Multiplying each fraction by a fraction of the appropriate form, and with unit value, produces an equivalent fraction whose denominator is the LCD. Therefore:

$$\frac{2x}{x^2-4}+\frac{3}{x^2-5x+6} = \frac{(x-3)(2x)}{(x-3)(x+2)(x-2)}+\frac{(x+2)(3)}{(x+2)(x-3)(x-2)}$$

$$= \frac{(x-3)(2x)+3(x+2)}{(x+2)(x-2)(x-3)}$$

distribute, $$= \frac{2x^2-6x+3x+6}{(x+2)(x-2)(x-3)}$$

combine terms $$= \frac{2x^2-3x+6}{(x+2)(x-2)(x-3)}.$$

Q Divide $\dfrac{2x-8}{x+1}$ by $\dfrac{3x^2-12x}{x^2-1}$.

A The problem can be written as:

$$\frac{\dfrac{2x-8}{x+1}}{\dfrac{3x^2-12x}{x^2-1}}.$$

To divide fractions invert the denominator and multiply the inverted fraction by the numerator. Thus,

$$\frac{\dfrac{2x-8}{x+1}}{\dfrac{3x^2-12x}{x^2-1}} = \frac{2x-8}{x+1} \times \frac{x^2-1}{3x^2-12x}.$$

Factor $$= \frac{2(x-4)}{x+1} \times \frac{(x+1)(x-1)}{3x(x-4)}.$$

Divide out common factors

$$= \frac{2(x-1)}{3x}.$$

 Divide $\frac{2y^2-11y+12}{6y^2-6y-12}$ by $\frac{3y^2-14y+8}{2y^2-6y+4}$.

Divide the two fractions by inverting the fraction to the right of the division sign and multiplying the inverted fraction by the fraction on the left of the division sign. Thus,

$$\frac{2y^2-11y+12}{6y^2-6y-12} \div \frac{3y^2-14y+8}{2y^2-6y+4}$$

$$= \frac{2y^2-11y+12}{6y^2-6y-12} \times \frac{2y^2-6y+4}{3y^2-14y+8}.$$

Now factor the expressions in the numerator and denominator of each fraction:

$$\frac{(2y-3)(y-4)2(y-2)(y-1)}{6(y-2)(y+1)(3y-2)(y-4)}.$$

Cancel common terms in the numerator and denominator to obtain:

$$\frac{(2y-3)(y-1)}{3(y+1)(3y-2)}.$$

CHAPTER 18

Complex Numbers

A complex number is an expression of the form $a + bi$, where a and b are real numbers and $i = \sqrt{-1}$. In the complex number, $a + bi$, a is called the real part and bi the imaginary part. The conjugate of a complex number $a + bi$ is $a - bi$.

Algebraic operations with complex numbers:

A) To add/subtract complex numbers, add/subtract the real and imaginary parts separately.

Example: $(a + bi) + (a - bi) = 2a$.

B) To multiply two complex numbers, treat the numbers as ordinary binomials and replace i^2 by -1.

Example: $(a + bi)(a - bi) = a^2 + abi - abi - (bi)^2 = a^2 + b^2$.

C) To divide two complex numbers, multiply the numerator and denominator of the fraction by the conjugate of the denominator, replacing i^2 by -1.

$$\frac{6+3i}{2+4i} = \frac{6+3i}{2+4i} \times \frac{2-4i}{2-4i}$$

$$= \frac{(6+3i)(2-4i)}{(2+4i)(2-4i)}$$

$$= \frac{12 + 6i - 24i - 12i^2}{4 + 8i - 8i - 16i^2}$$

$$= \frac{12 - 18i - 12(-1)}{4 - 16(-1)}$$

$$= \frac{12 - 18i + 12}{4 + 16}$$

$$= \frac{24 - 18i}{20}$$

$$= \frac{2(12 - 9i)}{2(10)}$$

$$= \frac{12 - 9i}{10}$$

$$= \frac{12}{10} - \frac{9}{10}i = \frac{6}{5} - \frac{9}{10}i$$

Complex numbers can be represented graphically. In Figure 18.1, the complex number $x + yi$ is represented graphically by the point P with the rectangular coordinates (x, y).

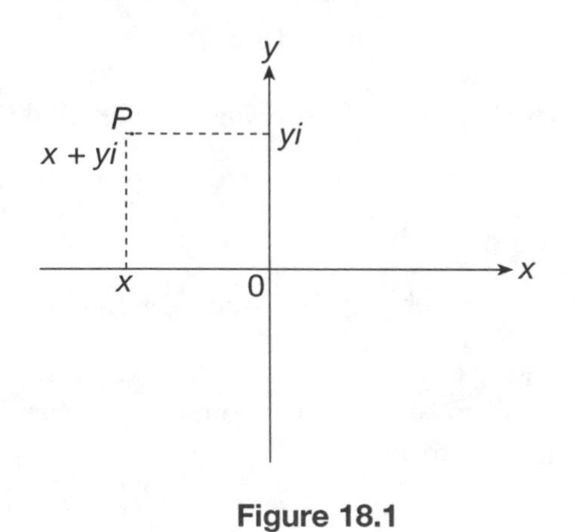

Figure 18.1

Here, the x-axis is called the real axis since all points on the x-axis have coordinates of the form $(x, 0)$ and correspond to real numbers $x + 0i = x$. Similarly, the y-axis is called the imaginary axis since all points on the y-axis correspond to pure imaginary numbers $0 + yi = yi$. The plane on which the complex numbers are represented is called the complex plane.

The complex numbers can also be represented by position vectors as illustrated in the Figure 18.2.

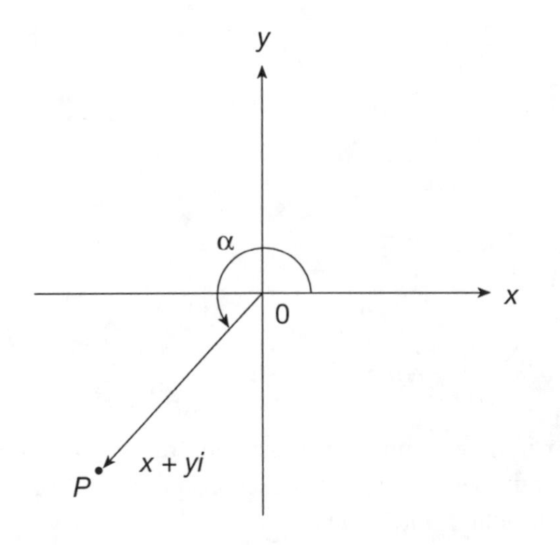

Figure 18.2

The vector in the above figure has a magnitude A equal to $\sqrt{x^2 + y^2}$, and the direction of this vector is given by an angle $\alpha = \tan^{-1}\frac{y}{x}$. Hence, we may express this complex number $x + yi$ as the vector $A \leq \alpha$, where $A = \sqrt{x^2 + y^2}$, $\alpha = \tan^{-1}\frac{y}{x}$, and $A \leq \mathrm{a} = x + yi$.

The complex number $N = A(\cos a + i\sin a)$ is said to be in trigonometric (or polar) form, whereas the complex number $N = x + yi$ is said to be in rectangular form.

18.1 DeMoivre's Theorem

The nth power of $A(\cos n\theta \; i\sin n\theta)$ is given by:

$$[A(\cos\theta + i\sin\theta)]^n = A^n(\cos n\theta + i\sin n\theta).$$

Problem Solving Examples:

Write each of the following in the form $a + bi$.

A) $(2 + 4i) + (3 + i)$

B) $(2 + i) - (4 - 2i)$

C) $(4 - i) - (6 - 2i)$

D) $3 - (4 + 2i)$

Solution:

A) $(2 + 4i) + (3 + i) = 2 + 4i + 3 + i$

$= (2 + 3) + 4i + i)$

$= 5 + 5i$

B) $(2 + i) - (4 - 2i) = 2 + i - 4 + 2i$

$= (2 - 4) + (i + 2i)$

$= -2 + 3i$

C) $(4 - i) - (6 - 2i) = 4 - i - 6 + 2i$

$= (4 - 6) + (-i + 2i)$

$= -2 + i$

D) $3 - (4 + 2i) = 3 - 4 - 2i$

$= (3 - 4) - 2i$

$= -1 - 2i$

 Simplify $\frac{3-5i}{2+3i}$.

A To simplify $\frac{3-5i}{2+3i}$ means to write the fraction without an imaginary number in the denominator. To achieve this, we multiply the fraction by another fraction which is equivalent to unity (so that the value of the original fraction is unchanged), which will transform the expression in the denominator to a real number. A fraction with this property must have the complex conjugate of the expression in the denominator of the original fraction as its numerator and denominator. The complex conjugate must be chosen because of its special property that when multiplied by the original complex number the result is real.

Note: $a + bi$; its complex conjugate is $a - bi$ or they can be said to be conjugates of each other. To multiply notice that $(a + bi)(a - bi)$ is the factored form of the difference of two squares. Thus, we obtain

$$\begin{aligned}(a)^2 - (bi)^2 &= (a)^2 - (-1)(b)^2, \text{ since } i^2 = -1 \\ &= a^2 + b^2.\end{aligned}$$

$$\begin{aligned}\frac{3-5i}{2+3i} \times \frac{2-3i}{2-3i} &= \frac{6-9i-10i+15i^2}{4-9i^2} \\ &= \frac{6-19i-15}{4+9} \\ &= \frac{-9-19i}{13} \text{ or } \frac{-9}{13} - \frac{19}{13}i.\end{aligned}$$

Since the resulting fraction has a rational number in the denominator, we have rationalized the denominator.

Quiz: Partial Fractions—Complex Numbers

1. If $i=\sqrt{-1}$, then $(a + bi)^2 - (a - bi)^2$ is equivalent to

   (A) $4abi$.
   (B) -1.
   (C) $a^2 - b^2$.
   (D) $2bi$.
   (E) $-2b^2$.

2. If $x = 3 + 2i$ and $y = 1 + 3i$, where $i^2 = -1$, then $\frac{x}{y} =$

   (A) $\frac{9}{10}-\frac{2}{3}i$.
   (B) $\frac{9}{10}-\frac{7}{10}i$.
   (C) $\frac{9}{10}+\frac{2}{3}i$.
   (D) $3-\frac{7}{10}i$.
   (E) $3+\frac{2}{3}i$.

3. If $x + yi = (3 + 2i)(1 + 3i)$, and $i=\sqrt{-1}$, then $x =$

   (A) -3.
   (B) -1.
   (C) 3.
   (D) 6.
   (E) 9.

4. Add $(3 + 4i)$ and $(2 - 5i)$.

   (A) $6 + 9i$
   (B) $5 - i$
   (C) $26 - 7\text{i}$
   (D) $6 - i$
   (E) $5 + 9i$

5. Subtract $7 - 2i$ from $-3 + 5i$.

(A) $10 - 7i$
(B) $-11 + 41i$
(C) $-5 + 2i$
(D) $-10 + 3i$
(E) $-10 + 7i$

6. Find the product $(2 + 3i)(-2 - 5i)$.

(A) $-11 + 16i$
(B) $-19 + 16i$
(C) -35
(D) $11 + 4i$
(E) $11 - 16i$

7. Which one of the following is the correct expression for the sum: $1^2 + 2^2 + 3^2 + \ldots + n^2$?

(A) $\frac{(n^2+1)(2n+1)}{6}$
(B) $\frac{(n+1)(2n+1)}{6}$
(C) $\frac{n(n+1)(2n+1)}{6}$
(D) $\frac{n^2 \times n}{6}$
(E) $n2.n$

8. Which one of the following is the correct expression for the sum of the first k terms of the series $1 + 5 + 9 + \ldots + (4k - 3)$?

(A) $k(2k - 1)$
(B) $(4k - 3)$
(C) $k(3k - 1)$
(D) $k^2 + 4$
(E) $k(2k + 1)$

9. Find the product $(-2 - 3i)(2 + 5i)$.

   (A) $11 - 16i$
   (B) -5
   (C) 5
   (D) $11 + 16i$
   (E) None of these.

10. Multiply $(3 + 4i)$ by $(5 + 2i)$.

   (A) $15 + 8i$
   (B) $15 + 8i^2$
   (C) $7 + 26i$
   (D) $7 + 26i^2$
   (E) $7 - 26i$

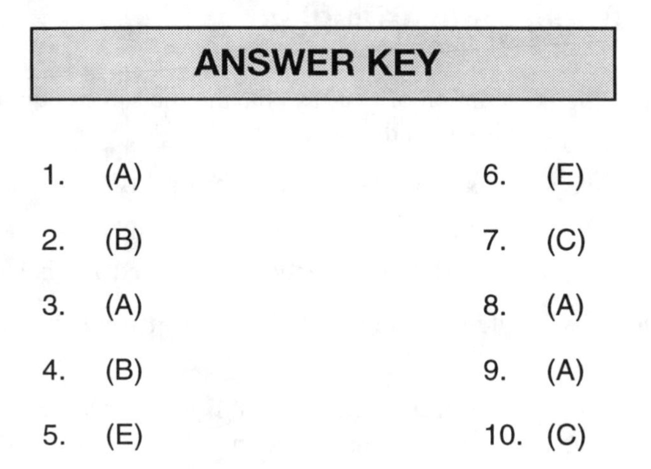

ANSWER KEY

1. (A)
2. (B)
3. (A)
4. (B)
5. (E)
6. (E)
7. (C)
8. (A)
9. (A)
10. (C)

CHAPTER 19

Trigonometry and Trigonometric Functions

19.1 Angles and Triangles

An angle is the union of two rays having the same endpoint. The common endpoint is called the vertex of the angle.

An acute angle is an angle that is larger than 0° but smaller than 90°.

An angle whose measure is exactly 90° is called a right angle.

An obtuse angle is an angle that is larger than 90° but less than 180°.

An angle whose measure is exactly 180° is called a straight angle. Note: such an angle is in fact a straight line.

An angle that is greater than 180° but less than 360° is called a reflex angle.

Two angles whose sum is 180° are called supplementary angles.

Two angles whose sum is 90° are called complementary angles.

Two angles are called adjacent angles if and only if they have a common vertex and a common side lying between them.

A pair of nonadjacent angles with a common vertex by two intersecting lines are called a pair of vertical angles. Vertical angles are equal.

Congruent angles are angles having equal measure.

A radian is defined as the measure of an angle, which when placed at the center of a circle, intercepts an arc of the circle equal to the radius of the circle.

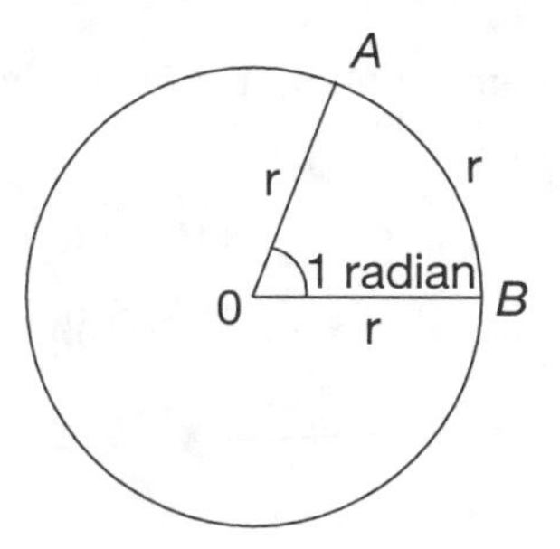

Figure 19.1

In Figure 19.1; $\angle AOB = 1$ radian.

$$1 \text{ radian} = \frac{180°}{\pi} \cong 57.3°$$

$$1° = \frac{\pi}{180°} \cong 0.0175 \text{ radian}$$

A closed three-sided geometric figure is called a triangle (Figure 19.2).

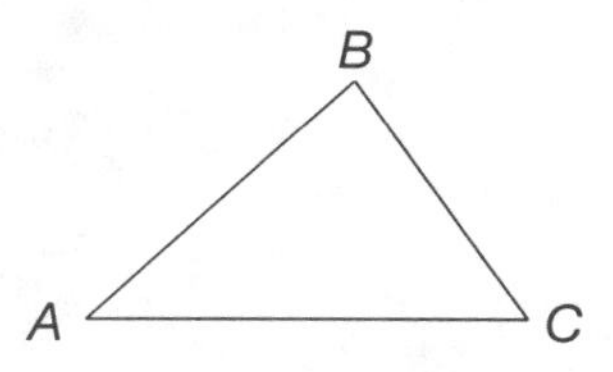

Figure 19.2

A triangle with one right angle is called a right triangle (Figure 19.3). The side opposite the right angle in a right triangle is called the hypotenuse of the right triangle. The other two sides are called the legs of the right triangle.

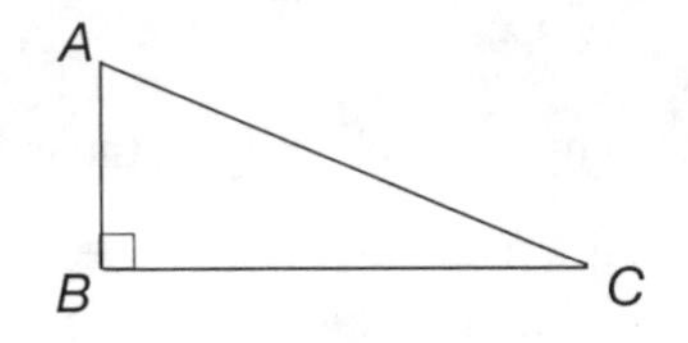

Figure 19.3

A triangle that does not contain a right angle is called an oblique triangle.

The sum of the interior angles of a triangle is 180°.

A triangle can have at most one right or obtuse angle.

If a triangle has two equal angles, then the sides opposite those angles are equal.

If two sides of a triangle are equal, then the angles opposite those sides are equal.

The sum of the exterior angles of a triangle, taking one angle at each vertex, is 360°.

A line that bisects one side of a triangle and is parallel to a second side, bisects the third side.

In a right triangle, the square of the hypotenuse is equal to the sum of the squares of the other two sides (Figure 19.4). This is commonly known as the theorem of Pythagoras or the Pythagorean theorem.

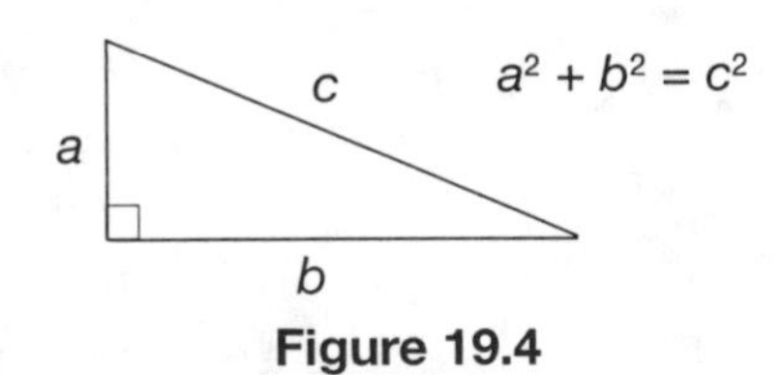

Figure 19.4

A portion of a circle is called an arc of the circle (Figure 19.5).

An angle whose vertex is at the center of a circle and whose sides are radii is called a central angle.

If α is the central angle in radians and r is the length of the radius of the circle, then the length of the intercepted arc l is given by $l = \alpha r$.

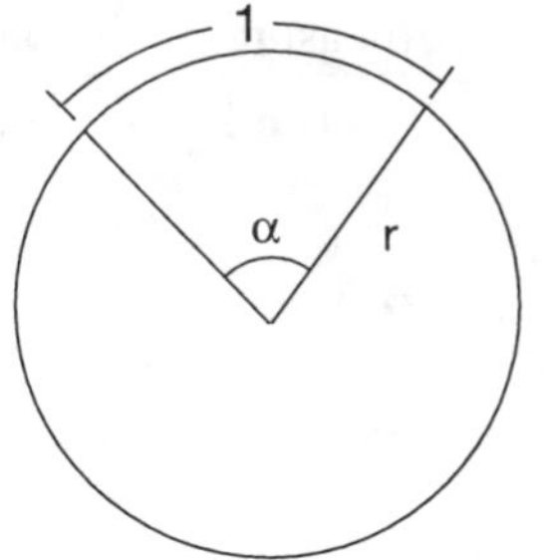

Figure 19.5

19.2 Trigonometric Ratios

Given a right triangle ΔABC as shown in Figure 19.6

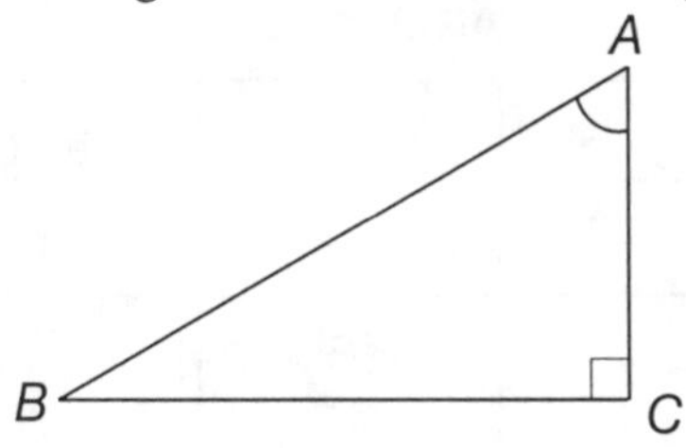

Figure 19.6

Definition 1:

$$\sin A = \frac{BC}{AB} = \frac{\text{measure of side opposite } \angle A}{\text{measure of hypotenuse}}$$

Definition 2:

$$\cos A = \frac{AC}{AB} = \frac{\text{measure of side adjacent to } \angle A}{\text{measure of hypotenuse}}$$

Definition 3: $\tan A = \dfrac{BC}{AC}$

$$= \frac{\text{measure of side opposite } \angle A}{\text{measure of side adjacent to } \angle A}$$

Definition 4: $\cot A = \dfrac{AC}{BC}$

$$= \frac{\text{measure of side adjacent to } \angle A}{\text{measure of side opposite to } \angle A}$$

$$\sec A = \frac{AB}{AC}$$

$$= \frac{\text{measure of hypotenuse}}{\text{measure of side adjacent to } \angle A}$$

$$\csc A = \frac{AB}{BC}$$

$$= \frac{\text{measure of hypotenuse}}{\text{measure of side opposite } \angle A}$$

Table 19.1 gives the values of sine, cosine, tangent, and cotangent for some special angles.

TABLE 19.1

α	Sin α	Cos α	Tan α	Cot α
$0°$	0	1	0	∞
$\frac{\pi^R}{6} = 30°$	$\frac{1}{2}$	$\frac{\sqrt{3}}{2}$	$\frac{\sqrt{3}}{3}$	$\sqrt{3}$
$\frac{\pi^R}{4} = 45°$	$\frac{\sqrt{2}}{2}$	$\frac{\sqrt{2}}{2}$	1	1
$\frac{\pi^R}{3} = 60°$	$\frac{\sqrt{3}}{2}$	$\frac{1}{2}$	$\sqrt{3}$	$\frac{\sqrt{3}}{3}$
$\frac{\pi^R}{2} = 90°$	1	0	∞	0

Problem Solving Examples:

Find tan 635°19′.

The reference angle of 635°19′ is 2(360°00′) – 635°19′ = 84°41′. Therefore, the tan 635°19′ = 84°41′ = 10.746. (This value may be found from a table of trigonometric functions.) However, the angle 635°19′ is a fourth quadrant angle and the tangent function is negative in the fourth quadrant. Hence, tan 635°19′ = –tan 84°41′ = –10.746.

Find sin 195°, cos 195°, tan 195°, and cot 195°.

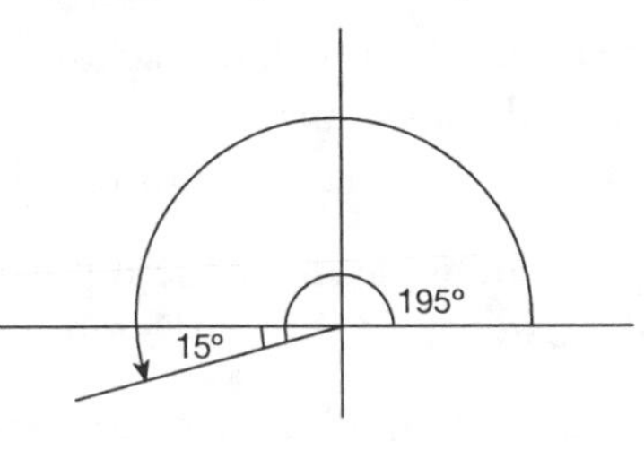

Figure 19.7

The reference angle for 195° is 15°. Also, 195° is a third quadrant angle (see Figure 19.7).

In the third quadrant, the sine and cosine functions are negative, but the tangent and the cotangent functions are positive. Therefore, sin 195° = –sin 15° = – 0.2588, cos 195° = – cos 15° = – 0.9659, tan 195° = tan 15° = .02679, and cot 195° = cot 15° = 3.7321. (Note that the values obtained for the trigonometric functions were found in a table of trigonometric functions.)

19.3 Trigonometric Functions

A circle with the center located at the origin of the rectangular coordinate axis and radius equal to one unit length is called a unit circle (see Figure 19.8).

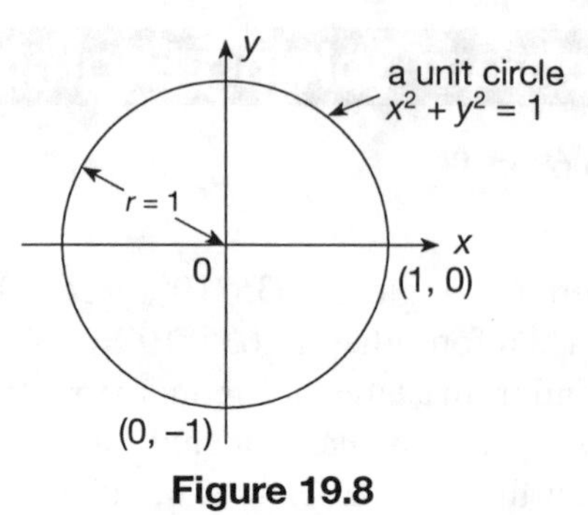

Figure 19.8

An angle whose vertex is at the origin of a rectangular coordinate system and whose initial side coincides with the positive x-axis is said to be in standard position with respect to the coordinate system.

An angle in standard position with respect to a Cartesian coordinate system whose terminal side lies in the first (or second or third or fourth) quadrant is called a first (or second or third or fourth) quadrant angle.

A quadrant angle is an angle in standard position whose terminal side lies on one of the axes of a Cartesian coordinate system.

If θ is a non-quadrantal angle in standard position and $P(x, y)$ is any point, distinct from the origin, on the terminal side of θ (Figures 19.9 and 19.10), then the six trigonometric functions of θ are defined in terms of the abscissa (x-coordinate), ordinate (y-coordinate), and distance $\overline{OP}$ as follows:

$$\text{sine } \theta = \sin\theta = \frac{\text{ordinate}}{\text{distance}} = \frac{y}{r}$$

$$\text{cosine } \theta = \cos\theta = \frac{\text{abscissa}}{\text{distance}} = \frac{x}{r}$$

$$\text{tangent } \theta = \tan\theta = \frac{\text{ordinate}}{\text{abscissa}} = \frac{y}{x}$$

$$\text{cotangent } \theta = \cot\theta = \frac{\text{abscissa}}{\text{ordinate}} = \frac{x}{y}$$

$$\text{secant } \theta = \sec\theta = \frac{\text{distance}}{\text{abscissa}} = \frac{r}{x}$$

$$\text{cosecant } \theta = \csc\theta = \frac{\text{distance}}{\text{ordinate}} = \frac{r}{y}$$

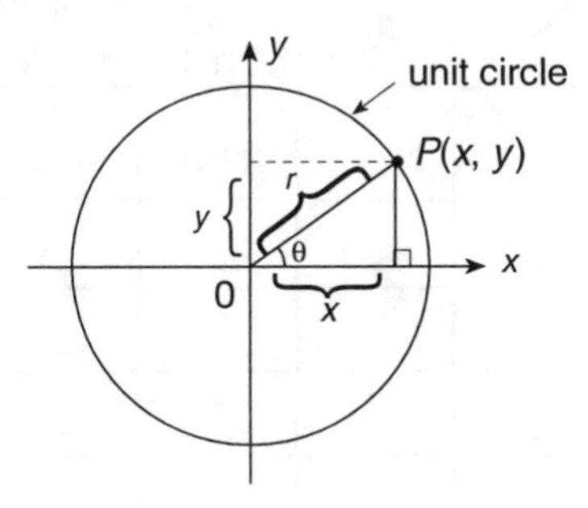

Figure 19.9

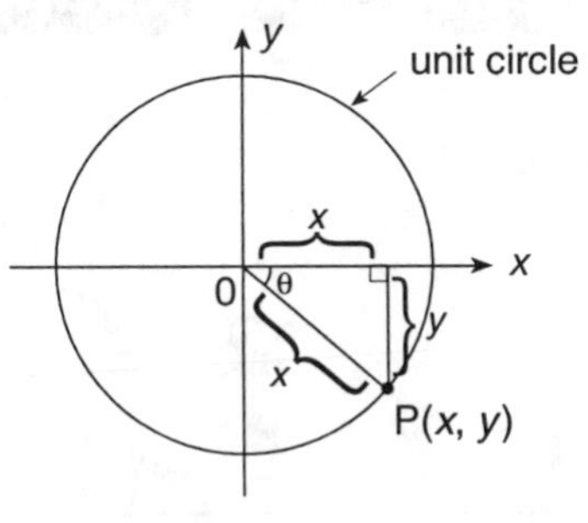

Figure 19.10

The value of trigonometric functions of quadrantal angles are given in Table 19.2.

TABLE 19.2

θ	sin θ	cos θ	tan θ	cot θ	sec θ	csc θ
0°	0	1	0	±∞	1	±∞
90°	1	0	±∞	0	±∞	1
180°	0	–1	0	±∞	–1	±∞
270°	–1	0	±∞	0	±∞	–1

Table 19.3 gives the signs of all the trigonometric functions for all four quadrants.

TABLE 19.3

Quadrant	sinα	cosα	tanα	cotα	secα	cscα
I	+	+	+	+	+	+
II	+	–	–	–	–	+
III	–	–	+	+	–	–
IV	–	+	–	–	+	–

Problem Solving Examples:

Find log sin 36°41′.

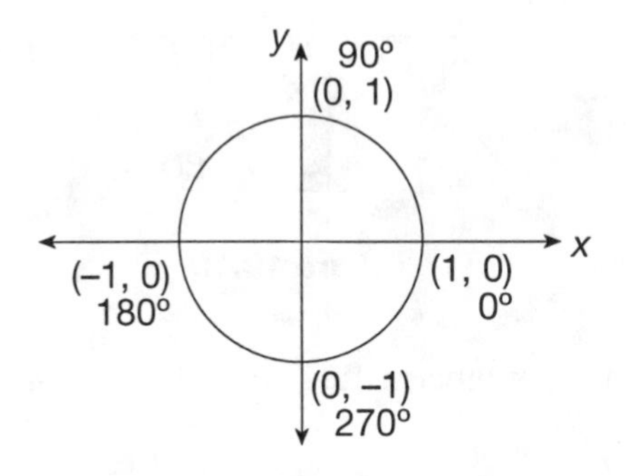

Figure 19.11

A From a table of logarithms of trigonometric functions, we find log sin 36°40′ and log sin 36°50′. Then, by the process of interpolation, we find: log sin 36°41′ = 9.77626 – 10. Since all values of the sine function for acute angles are in the range of $0 \leq \sin x \leq 1$, the characteristic is negative. (Recall that for a number less than one, the characteristic is negative.) The range of sine can be seen by inspecting Figure 19.11. Sine is given by the y coordinate; cos is given by the x coordinate. Observe that the y value varies from 0 to 1, as the angle varies from 0° to 90°.

Q Find log cos 49°13.6′.

First consult a table of logarithms of trigonometric functions.

Notice that 49°13.6′ lies between 49°10′ and 49°20′, so that the log of 49°13.6′ will occur between the logs of 49°10′ and 49°20′ and can be determined by interpolation.

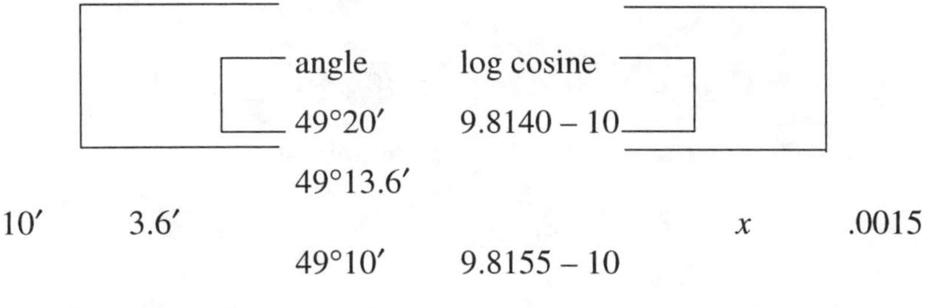

Now set up the proportion

$$\frac{3.6}{10} = \frac{x}{.0015} \qquad \text{or} \qquad 10x = .00540$$

$$x = .000540$$

Since cos decreases on the interval $0 < \theta < \pi/2$, subtract x from the log cosine of 49°10′.

$$\begin{array}{r} 9.8155 \\ -\quad .000540 \\ \hline 9.814960 \end{array}$$

Thus, log cosine 49°13.6′ is 9.81496 – 10.

19.4 Properties and Graphs of Trigonometric Functions; Trigonometric Identities and Formulas

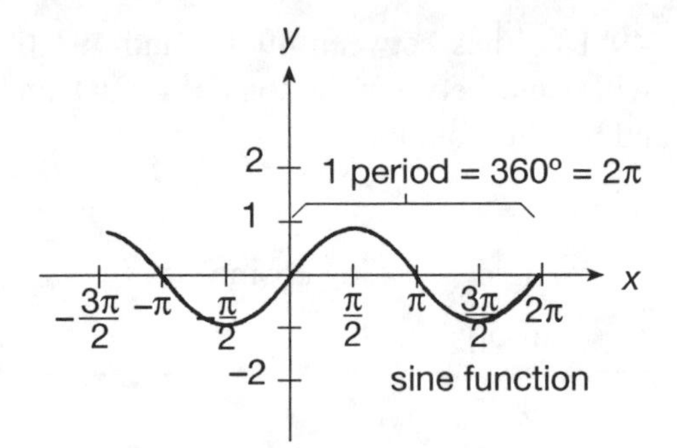

Figure 19.12

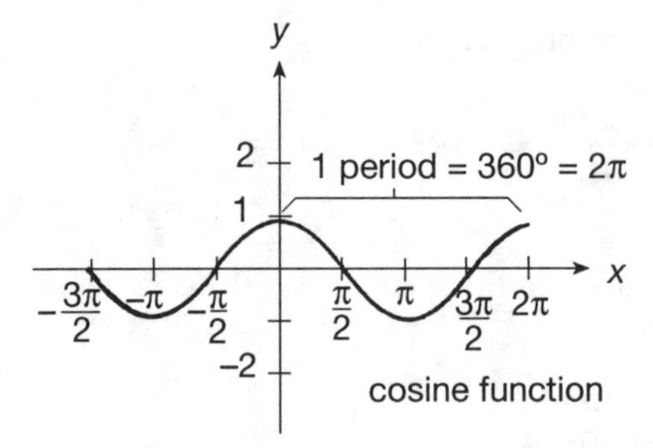

Figure 19.13

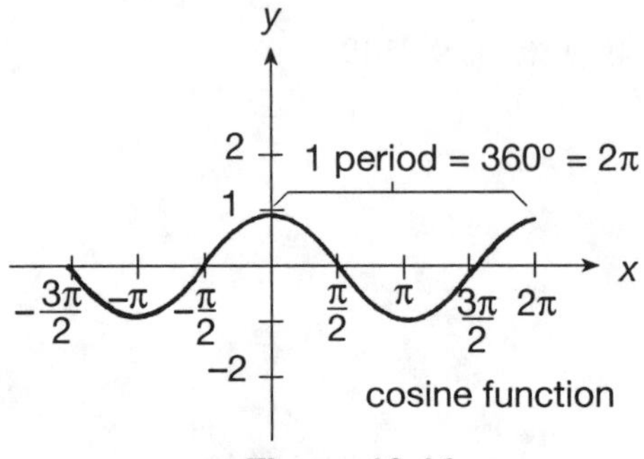

Figure 19.14

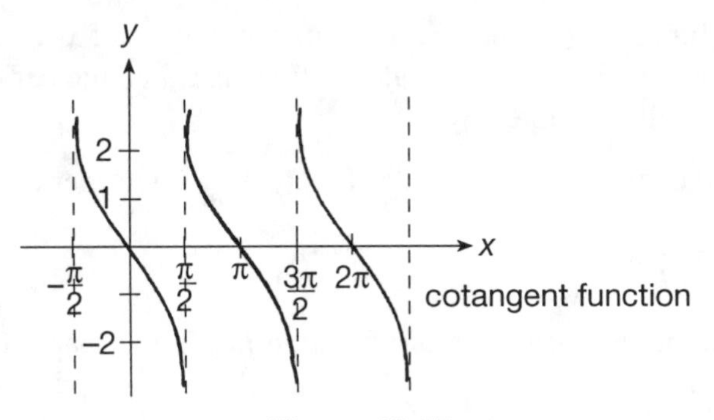

Figure 19.15

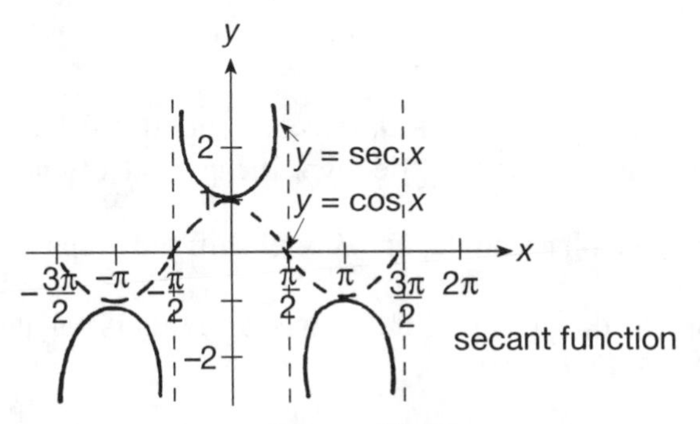

Figure 19.16

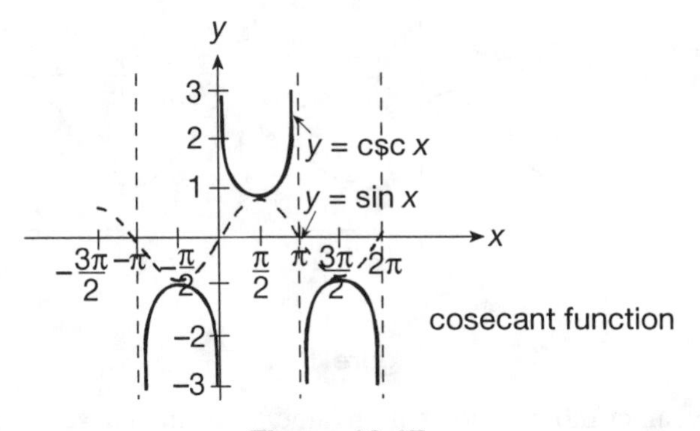

Figure 19.17

A periodic function is defined as a function that repeats its values in definite cycles (Figure 19.12, 19.13, and 19.14). Trigonometric functions are periodic functions (Figures 19.15, 19.16, and 19.17).

For a function $f(x)$, the smallest range of values of x which corresponds to a complete cycle of values of $f(x)$ is called the period of the function, and it is denoted by T.

The frequency f of a periodic function $f(x)$ with period T is frequency $f = \frac{1}{T}$.

The amplitude of a periodic function $f(x)$ is the maximum value of its ordinate.

For a periodic function $f(x) = \sin(ax + \phi)$, ϕ is called the phase angle and $-\frac{\phi}{a}$ is called the phase shift.

Given $y = A \sin(Bx + C) + D$, the constant $|A|$ is the amplitude of the function, the constant B decides the period of the function, $T = \frac{2\pi}{|B|}$. C is the phase angle, and D will shift the graph of $y = A \sin(Bx + C)$ up (or down) along the y-axis by D units for positive (or negative) D (see Figure 19.18).

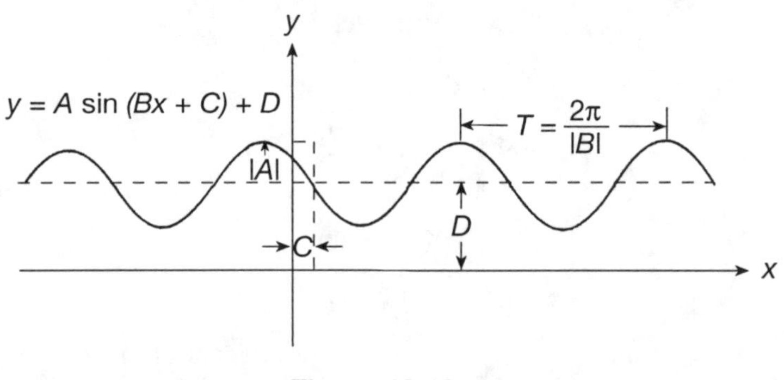

Figure 19.18

One can combine two or more sine or cosine curves as illustrated by the following example to obtain more complicated forms of wave motions.

Graph (1) $y = 2 \sin x + 4 \sin x$

(2) $y = \sin x + \cos x$

(1) To obtain the graph of $y = 2 \sin x + 4 \sin x$, we first draw the graphs of $y_1 = 2 \sin x$ and $y_2 = 4 \sin x$ as shown in Figure 19.19. Next, draw the graph of $y = y_1 + y_2$ by adding corresponding ordinates as illustrated in Figure 19.19. For instance, at $x = a_1$, the ordinate of y is a_1d_1 which is the algebraic sum of the ordinates a_1b_1 of y_1 and a_1c_1 of y_2.

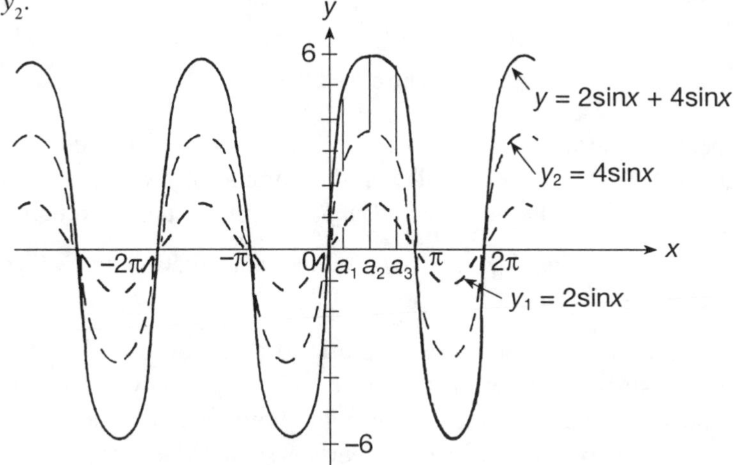

Figure 19.19

Another way of obtaining the graph of the given function is to construct a table such as Table 19.4.

TABLE 19.4

x	...	$-\frac{\pi}{2}$	$-\frac{\pi}{3}$	$-\frac{\pi}{4}$	$-\frac{\pi}{6}$	0	$\frac{\pi}{6}$	$\frac{\pi}{4}$	$\frac{\pi}{3}$	$\frac{\pi}{2}$	...
$y = 2\sin x + 4\sin x$	...	-6	$-3\sqrt{3}$	$-3\sqrt{2}$	-3	0	3	$3\sqrt{2}$	$3\sqrt{3}$	6	...

(2) The graph of $y = \sin x + \cos x$ is obtained by using the methods described in part (1); it is shown in Figure 19.20.

A function $f(x)$ is said to be an even function if $f(-x) = f(x)$ for all

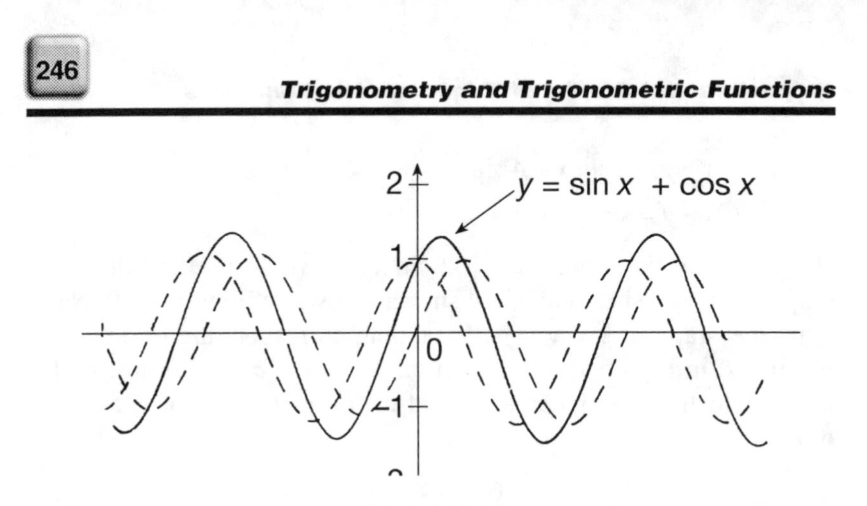

Figure 19.20

independent variables x in the domain of f. $f(x)$ is said to be an odd function if $f(-x) = -f(x)$ for all x in the domain of f. Since for $f(x) = \sin x$, $f(-x) = \sin(-x) = -\sin x = -f(x)$, $f(x) = \sin x$ is an odd function.

On the other hand, $f(x) = \cos x$ is an even function since $f(-x) = \cos(-x) = \cos x = f(x)$.

A curve C is said to be asymptotic to a straight line l if (a) the shortest distance d between a point on C and l is never zero, (b) d approaches zero as either x or y or both coordinates of P approaches ∞ or $-\infty$. For example, one can clearly see that the curve of $f(x) = \tan x$ (Figure 19.14) has the property of an asymptotic function.

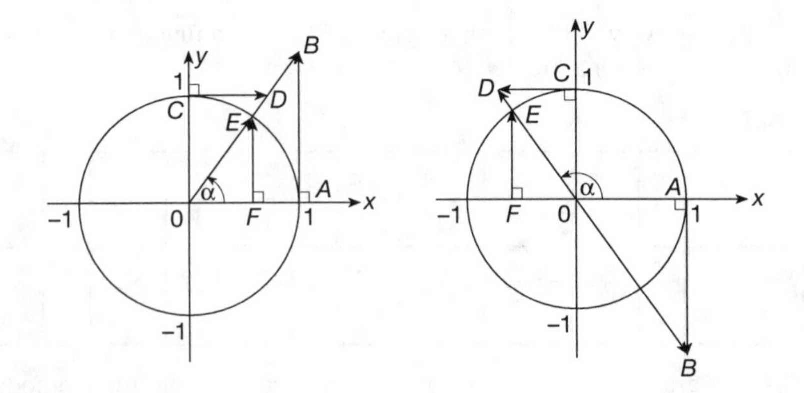

Figure 19.21

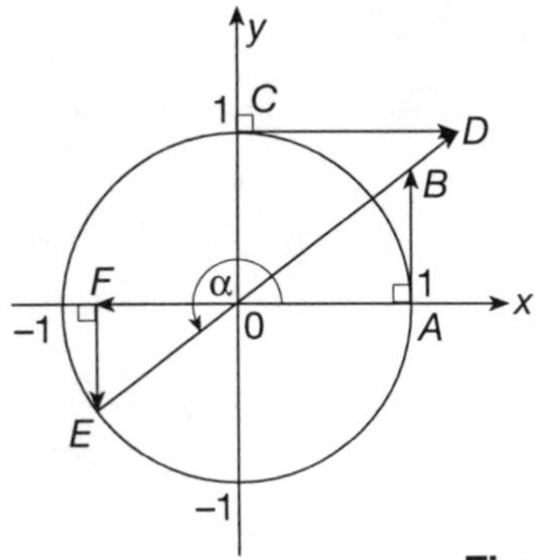

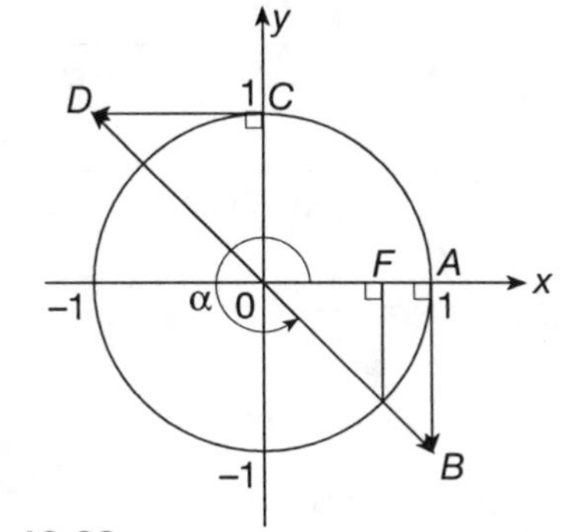

Figure 19.22

$$\sin \alpha = \frac{FE}{OE} = FE$$

$$\cos \alpha = \frac{OF}{OE} = OF$$

$$\tan \alpha = \frac{FE}{OF} = \frac{AB}{OA} = AB$$

$$\cot \alpha = \frac{OF}{FE} = \frac{CD}{OC} = CD$$

$$\sec \alpha = \frac{OE}{OF} = \frac{OB}{OA} = OB$$

$$\csc \alpha = \frac{OE}{FE} = \frac{OD}{OC} = OD$$

We can represent the values of the trigonometric functions as the lengths of the line segments in a unit circle as shown in Figures 19.21 and 19.22.

19.4.1 Fundamental Relations and Identities

$$\sin^2 \alpha + \cos^2 \alpha = 1$$

$$\tan \alpha = \frac{\sin \alpha}{\cos \alpha}$$

$$\cot\alpha = \frac{\cos\alpha}{\sin\alpha} = \frac{1}{\tan\alpha}$$

$$\csc\alpha = \frac{1}{\sin\alpha}$$

$$\sec\alpha = \frac{1}{\cos\alpha}$$

$$1 + \tan^2\alpha = \sec^2\alpha$$

$$1 + \cot^2\alpha = \csc^2\alpha$$

One can find all the trigonometric functions of an acute angle when the value of any one of them is known.

For example, given α is an acute angle and $\csc\alpha = 2$.

Then:

$$\sin\alpha = \frac{1}{\csc\alpha} = \frac{1}{2}$$

$$\cos^2\alpha + \sin^2\alpha = 1 \qquad \cos\alpha = \sqrt{1-\sin^2\alpha}$$

$$= \sqrt{1-\left(\tfrac{1}{2}\right)^2}$$

$$= \sqrt{1-\tfrac{1}{4}}$$

$$= \frac{\sqrt{3}}{2}$$

$$\tan\alpha = \frac{\sin\alpha}{\cos\alpha} = \frac{\frac{1}{2}}{\frac{\sqrt{3}}{2}} = \frac{1}{\sqrt{3}} = \frac{\sqrt{3}}{3}$$

$$\cot\alpha = \frac{1}{\tan\alpha} = \sqrt{3}$$

$$\sec\alpha = \frac{1}{\cos\alpha} = \frac{1}{\frac{\sqrt{3}}{2}} = \frac{2}{\sqrt{3}} = \frac{2\sqrt{3}}{3}$$

For a given angle θ in standard position, the related angle of θ is the unique acute angle which the terminal side of θ makes with the *x*-axis (Figures 19.23, 19.24, and 19.25).

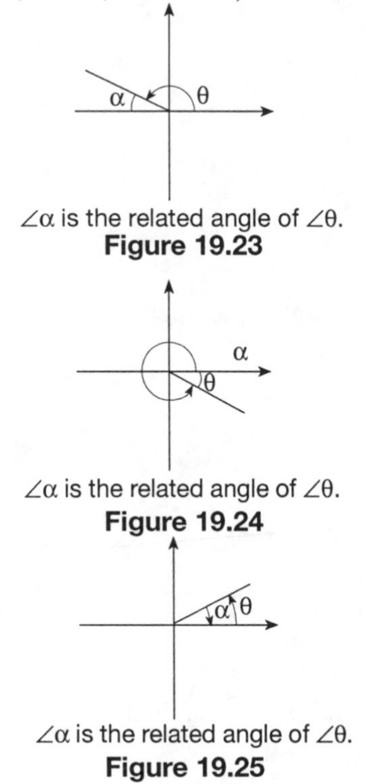

∠α is the related angle of ∠θ.
Figure 19.23

∠α is the related angle of ∠θ.
Figure 19.24

∠α is the related angle of ∠θ.
Figure 19.25

Let θ be an angle in standard position and ϕ be the related angle of θ.

A) If θ is a first quadrant angle, then

(a) $\sin\theta = \sin\phi$

(b) $\cos\theta = \cos\phi$

(c) $\tan\theta = \tan\phi$

(d) $\cot\theta = \cot\phi$

(e) $\sec\theta = \sec\phi$

(f) $\csc\theta = \csc\phi$

B) If θ is a second quadrant angle, then

(a) $\sin\theta = \sin\phi$

(b) $\cos\theta = -\cos\phi$

(c) $\tan\theta = -\tan\phi$

(d) $\cot\theta = -\cot\phi$

(e) $\sec\theta = -\sec\phi$

(f) $\csc\theta = \csc\phi$

C) If θ is a third quadrant angle, then

(a) $\sin\theta = -\sin\phi$

(b) $\cos\theta = -\cos\phi$

(c) $\tan\theta = \tan\phi$

(d) $\cot\theta = \cot\phi$

(e) $\sec\theta = -\sec\phi$

(f) $\csc\theta = -\csc\phi$

D) If θ is a fourth quadrant angle, then

(a) $\sin\theta = -\sin\phi$

(b) $\cos\theta = \cos\phi$

(c) $\tan\theta = -\tan\phi$

(d) $\cot\theta = -\cot\phi$

(e) $\sec\theta = \sec\phi$

(f) $\csc\theta = -\csc\phi$

19.4.2 Addition and Subtraction Formulas

Use Table 19.5 to solve the following formulas:

$$\sin(A \pm B) = \sin A \cos B \pm \cos A \sin B$$

$$\cos(A \pm B) = \cos A \cos B \pm \sin A \sin B$$

$$\tan(A \pm B) = \frac{\tan A \pm \tan B}{1 \pm \tan A \tan B}$$

$$\cot(A \pm B) = \frac{\cot A \cot B \pm 1}{\cot B \pm \cot A}$$

TABLE 19.5

	sin	cos	tan	cot	sec	csc
$-\alpha$	$-\sin\alpha$	$+\cos\alpha$	$-\tan\alpha$	$-\cot\alpha$	$+\sec\alpha$	$-\csc\alpha$
$90° + \alpha$	$+\cos\alpha$	$-\sin\alpha$	$-\cot\alpha$	$-\tan\alpha$	$-\csc\alpha$	$+\sec\alpha$
$90° - \alpha$	$+\cos\alpha$	$+\sin\alpha$	$+\cot\alpha$	$+\tan\alpha$	$+\csc\alpha$	$+\sec\alpha$
$180° + \alpha$	$-\sin\alpha$	$-\cos\alpha$	$+\tan\alpha$	$+\cot\alpha$	$-\sec\alpha$	$-\csc\alpha$
$180° - \alpha$	$+\sin\alpha$	$-\cos\alpha$	$-\tan\alpha$	$-\cot\alpha$	$-\sec\alpha$	$+\csc\alpha$
$270° + \alpha$	$-\cos\alpha$	$+\sin\alpha$	$-\cot\alpha$	$-\tan\alpha$	$+\csc\alpha$	$-\sec\alpha$
$270° - \alpha$	$-\cos\alpha$	$-\sin\alpha$	$+\cot\alpha$	$+\tan\alpha$	$-\csc\alpha$	$-\sec\alpha$
$360° + \alpha$	$+\sin\alpha$	$+\cos\alpha$	$+\tan\alpha$	$+\cot\alpha$	$+\sec\alpha$	$+\csc\alpha$
$360° - \alpha$	$-\sin\alpha$	$+\cos\alpha$	$-\tan\alpha$	$-\cot\alpha$	$+\sec\alpha$	$-\csc\alpha$

19.4.3 Double-angle Formulas

$$\sin 2A = 2\sin A\cos A$$

$$\cos 2A = 2\cos^2 A - 1$$

$$= 1 - 2\sin^2 A$$

$$= \cos^2 A - \sin^2 A$$

$$\tan 2A = \frac{2\tan A}{1 - \tan^2 A}$$

19.4.4 Half-angle Formulas

$$\sin\frac{A}{2} = \pm\sqrt{\frac{1-\cos A}{2}}$$

$$\cos\frac{A}{2} = \pm\sqrt{\frac{1+\cos A}{2}}$$

$$\tan\frac{A}{2} = \pm\sqrt{\frac{1-\cos A}{1+\cos A}}$$

$$= \frac{1-\cos A}{\sin A}$$

$$= \frac{\sin A}{1+\cos A}$$

$$\cot\frac{A}{2} = \pm\sqrt{\frac{1+\cos A}{1-\cos A}} = \frac{1+\cos A}{\sin A} = \frac{\sin A}{1-\cos A}$$

19.4.5 Sum and Difference Formulas

$$\sin\alpha + \sin\beta = 2\sin\left(\frac{\alpha+\beta}{2}\right)\cos\left(\frac{\alpha-\beta}{2}\right)$$

$$\sin\alpha - \sin\beta = 2\cos\left(\frac{\alpha+\beta}{2}\right)\sin\left(\frac{\alpha-\beta}{2}\right)$$

$$\cos\alpha + \cos\beta = 2\cos\left(\frac{\alpha+\beta}{2}\right)\cos\left(\frac{\alpha-\beta}{2}\right)$$

$$\cos\alpha - \cos\beta = -2\sin\left(\frac{\alpha+\beta}{2}\right)\sin\left(\frac{\alpha-\beta}{2}\right)$$

$$\tan\alpha + \tan\beta = \frac{\sin(\alpha+\beta)}{\cos\alpha\cos\beta}$$

$$\tan\alpha - \tan\beta = \frac{\sin(\alpha-\beta)}{\cos\alpha\cos\beta}$$

19.4.6 Product Formulas of Sines and Cosines

$$\sin A \sin B = \frac{1}{2}[\cos(A-B) - \cos(A+B)]$$

$$\cos A \cos B = \frac{1}{2}[\cos(A+B) + \cos(A-B)]$$

$$\sin A \cos B = \frac{1}{2}[\sin(A+B) + \sin(A-B)]$$

$$\cos A \sin B = \frac{1}{2}[\sin(A+B) - \sin(A-B)]$$

Problem Solving Examples:

Q Change $4 + (\tan\theta - \cot\theta)^2$ to $\sec^2\theta + \csc^2\theta$.

A If we square the binomial in the first expression, we have

$$\begin{aligned} 4 + (\tan\theta - \cot\theta)^2 &= 4 + (\tan\theta - \cot\theta)(\tan\theta - \cot\theta) \\ &= 4 + \tan^2\theta - 2\tan\theta\cot\theta + \cot^2\theta. \end{aligned}$$

Since $\cot\theta = \dfrac{1}{\tan\theta}$, the term $-2\tan\theta\cot\theta = -2\tan\theta\left(\dfrac{1}{\tan\theta}\right) = -2(1) = -2$.

Thus, $4 + (\tan\theta - \cot\theta)^2 = 4 + \tan^2\theta - 2 + \cot^2\theta$

$= 2 + \tan^2\theta + \cot^2\theta$

Since $2 = 1 + 1$, $\quad = 1 + \tan^2\theta + 1 + \cot^2\theta$.

Recall $1 + \tan^2\theta = \sec^2\theta$ and $1 + \cot^2\theta = \csc^2\theta$. Replacing these values, we obtain

$$4 + (\tan\theta - \cot\theta)^2 = \sec^2\theta + \csc^2\theta.$$

Reduce the expression $\dfrac{\tan x - \cot x}{\tan x + \cot x}$ to one involving only $\sin x$.

Since, by definition, $\tan x = \dfrac{\sin x}{\cos x}$ and

$$\cot x = \frac{1}{\tan x} = \frac{1}{\sin x / \cos x} = \frac{\cos x}{\sin x},$$

$$\begin{aligned}
\frac{\tan x - \cot x}{\tan x + \cot x} &= \frac{\dfrac{\sin x}{\cos x} - \dfrac{\cos x}{\sin x}}{\dfrac{\sin x}{\cos x} + \dfrac{\cos x}{\sin x}} \\
&= \frac{\dfrac{\sin x(\sin x)}{\sin x(\cos x)} - \dfrac{\cos x(\cos x)}{\cos x(\sin x)}}{\dfrac{\sin x(\sin x)}{\sin x(\cos x)} + \dfrac{\cos x(\cos x)}{\cos x(\sin x)}} \\
&= \frac{\dfrac{\sin^2 x - \cos^2 x}{\sin x \cos x}}{\dfrac{\sin^2 x + \cos^2 x}{\sin x \cos x}} \\
&= \frac{\sin^2 x - \cos^2 x}{\sin x \cos x} \times \frac{\sin x \cos x}{\sin^2 x + \cos^2 x} \\
&= \frac{\sin^2 x - \cos^2 x}{\sin^2 x + \cos^2 x}.
\end{aligned}$$

Since $\sin^2 x + \cos^2 x = 1$ or $\cos^2 x = 1 - \sin^2 x$,

$$\begin{aligned}
\frac{\tan x - \cot x}{\tan x + \cot x} &= \frac{\sin^2 x - \cos^2 x}{\sin^2 x + \cos^2 x} = \frac{\sin^2 x - \cos^2 x}{1} \\
&= \sin^2 x - \cos^2 x \\
&= \sin^2 x - (1 - \sin^2 x) \\
&= \sin^2 x - 1 + \sin^2 x \\
&= 2 \sin^2 x - 1.
\end{aligned}$$

CHAPTER 20

Solving Triangles

20.1 Laws and Formulas

The following results hold for any oblique triangle *ABC* with sides of length *a*, *b*, *c* opposite to vertices *A*, *B*, *C*, respectively.

20.1.1 Law of Sines

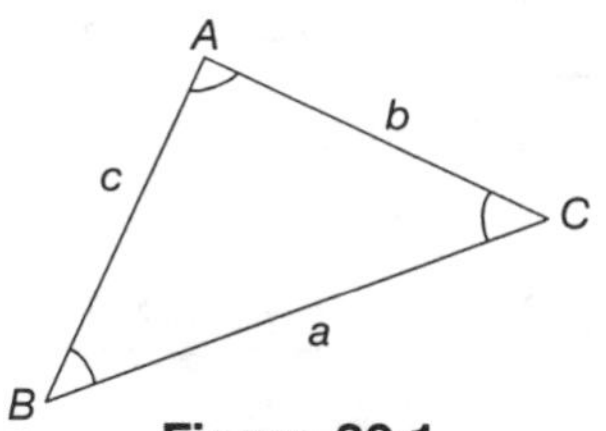

Figure 20.1

$$\frac{a}{\sin A} = \frac{b}{\sin B} = \frac{c}{\sin C}$$

20.1.2 Law of Cosines

$$a^2 = b^2 + c^2 - 2bc \cos A$$

$$b^2 = a^2 + c^2 - 2ac \cos B$$

$$c^2 = a^2 + b^2 - 2ab \cos C$$

20.1.3 Law of Tangents

$$\frac{a-b}{a+b}=\frac{\tan\left(\frac{A-B}{2}\right)}{\tan\left(\frac{A+B}{2}\right)}$$

$$\frac{b-c}{b+c}=\frac{\tan\left(\frac{B-C}{2}\right)}{\tan\left(\frac{B+C}{2}\right)}$$

$$\frac{a-c}{a+c}=\frac{\tan\left(\frac{A-C}{2}\right)}{\tan\left(\frac{A+C}{2}\right)}$$

20.1.4 Mollweide's Formulas

$$\frac{a+b}{c}=\frac{\cos\frac{1}{2}(A-B)}{\sin\frac{C}{2}} \qquad \frac{a-b}{c}=\frac{\sin\frac{1}{2}(A-B)}{\cos\frac{C}{2}}$$

$$\frac{b+c}{a}=\frac{\cos\frac{1}{2}(B-C)}{\sin\frac{A}{2}} \qquad \frac{b-c}{a}=\frac{\sin\frac{1}{2}(B-C)}{\cos\frac{A}{2}}$$

$$\frac{c+a}{b}=\frac{\cos\frac{1}{2}(C-A)}{\sin\frac{A}{2}} \qquad \frac{c-a}{b}=\frac{\sin\frac{1}{2}(C-A)}{\cos\frac{B}{2}}$$

20.1.5 Projection Formulas

$$BC = b\cos C + c\cos B$$

$$AC = a\cos C + c\cos A$$

$$AB = a\cos B + b\cos A$$

In a 30°– 60° right triangle (Figure 20.2), the hypotenuse is twice the length of the side opposite the 30° angle. The side opposite the 60° angle is equal to the length of the side opposite the 30° angle multiplied by $\sqrt{3}$.

In an isosceles 45° right triangle (Figure 20. 3), the hypotenuse is equal to the length of one of its arms multiplied by $\sqrt{2}$.

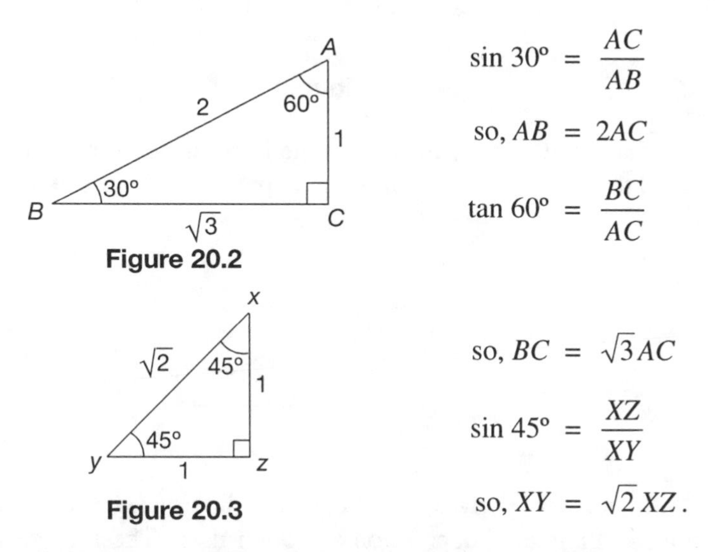

Figure 20.2

Figure 20.3

20.2 Important Concepts and Theorems

The altitude h on the hypotenuse of a right triangle is the mean proportional between the segments of the hypotenuse, also called the geometric mean (Figure 20.4).

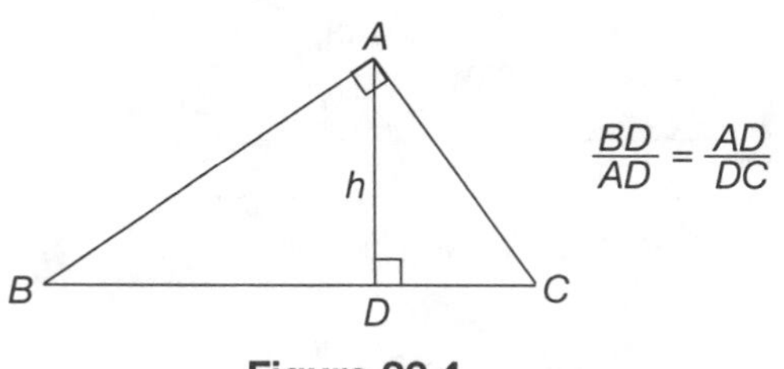

Figure 20.4

In a right triangle ΔABC, the altitude to the hypotenuse, $\overline{AD}$, separates the triangle into two triangles that are similar to each other and to the original triangle (Figure 20.5).

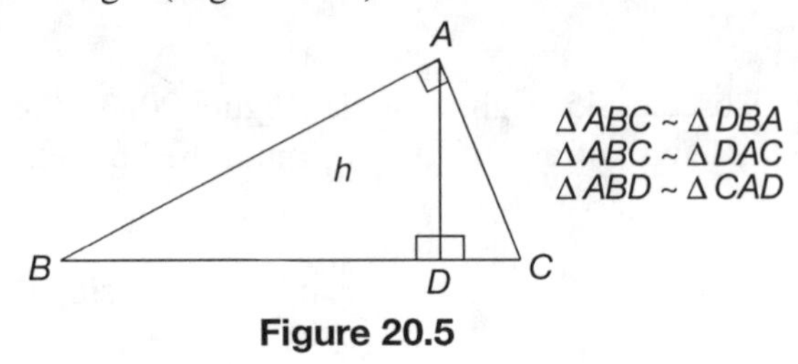

Figure 20.5

The length of the median to the hypotenuse of a right triangle is equal to one-half the length of the hypotenuse (Figure 20.6).

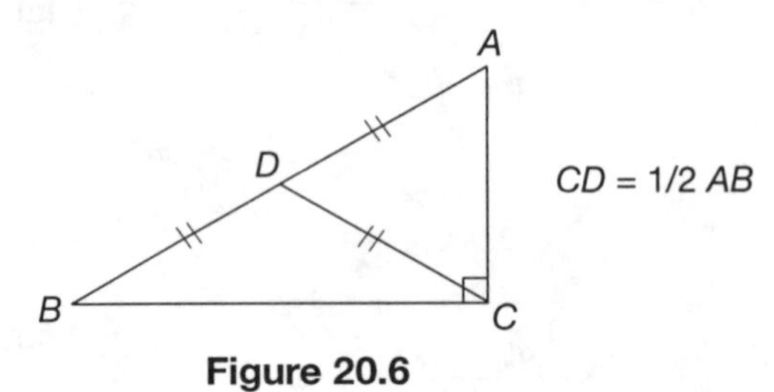

Figure 20.6

In solving triangles, terms such as line of sight, angle of elevation, and angle of depression are often used. These terms are illustrated in Figure 20.7 and 20.8.

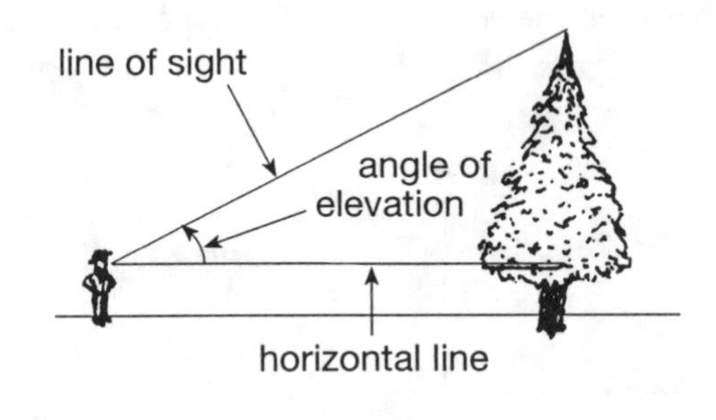

Figure 20.7

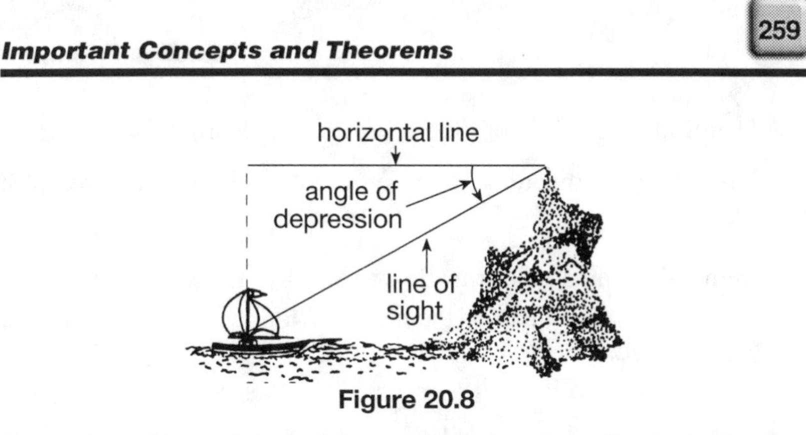

Figure 20.8

Example: At a point on the ground 40 feet from the foot of a tree, the angle of elevation to the top of the tree is 42°. Find the height of the tree to the nearest foot.

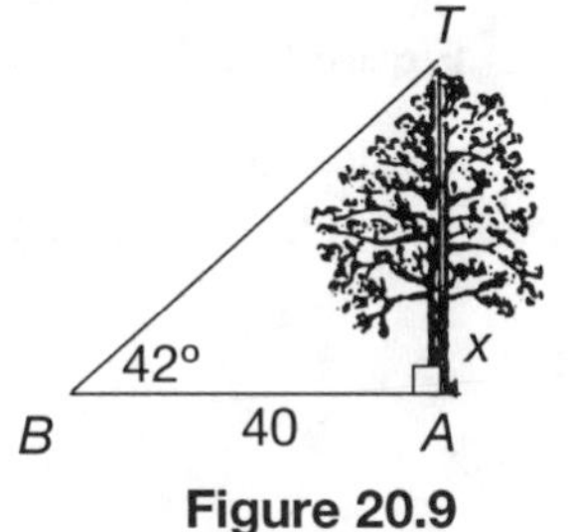

Figure 20.9

Solution: The geometric figure formed is a right triangle (see Figure 20.9). Since the unknown height of the tree is opposite the given angle of elevation, and we are given the side adjacent to this angle, we can solve the problem using the tangent ratio. The tangent is the ratio of the length of the leg opposite the acute angle to the length of the leg adjacent to the acute angle in any right triangle. In this example,

$$\tan B = \frac{\text{length of leg opposite } \angle B}{\text{length of leg adjacent } \angle B}$$

$$\tan B = \frac{AT}{BA}.$$

Let $x = AT$, and consult a standard table of tangents to find that $\tan 42° = 0.9004$. Since $BA = 40$, we obtain

$$0.9004 = \frac{x}{40}.$$

Therefore, $x = 40(0.9004) = 36.016.$

Therefore, the height of the tree, to the nearest foot, is 36 feet.

The following theorems are often used for finding the area of a triangle.

Theorem: The area of a triangle is given by $A = \frac{1}{2}bh$, where b is the length of the base and h is the perpendicular height of the triangle (Figure 20.10).

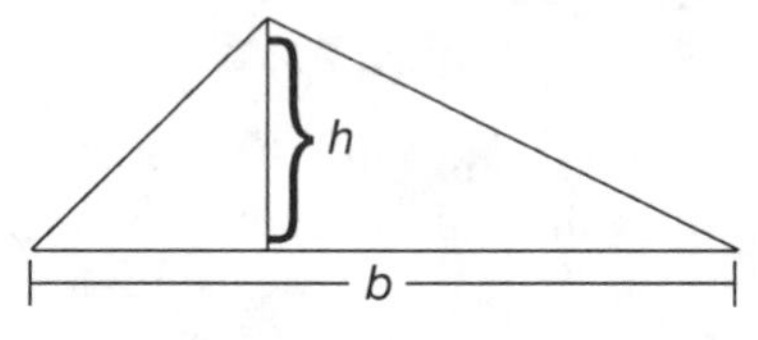

Figure 20.10

Theorem: The area of a triangle equals one-half the product of any two adjacent sides and the sine of the included angle (Figure 20.11).

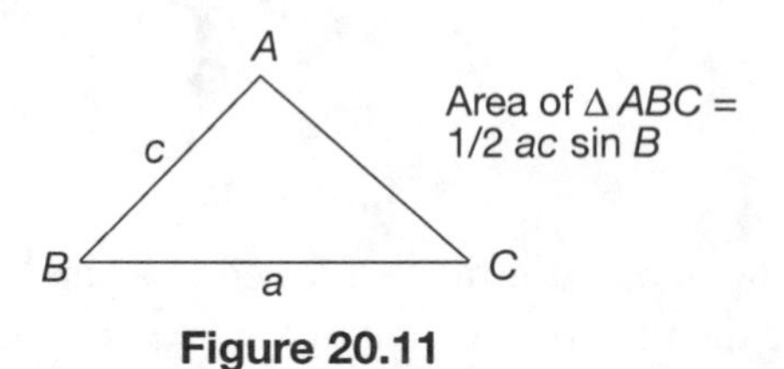

Figure 20.11

Theorem: Triangles that share the same base and have their third vertex on a line parallel to the base have equal areas (Figure 20.12).

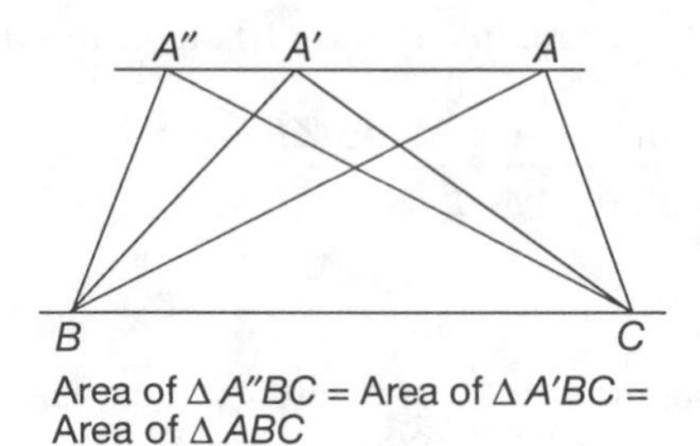

Figure 20.12

The areas of two triangles having equal bases have the same ratio as that of their altitudes and vice versa.

The area of a triangle, the length of whose three sides are a, b, and c, is given by the formula

$$A = \sqrt{s(s-a)(s-b)(s-c)}$$

where $s = \frac{1}{2}(a+b+c)$; the semiperimeter of the triangle. The above formula is commonly referred to as Heron's formula.

The area of an equilateral triangle is given by the formula,

$$A = \frac{x^2\sqrt{3}}{4},$$

where x is the length of a side of the triangle (Figure 20.13).

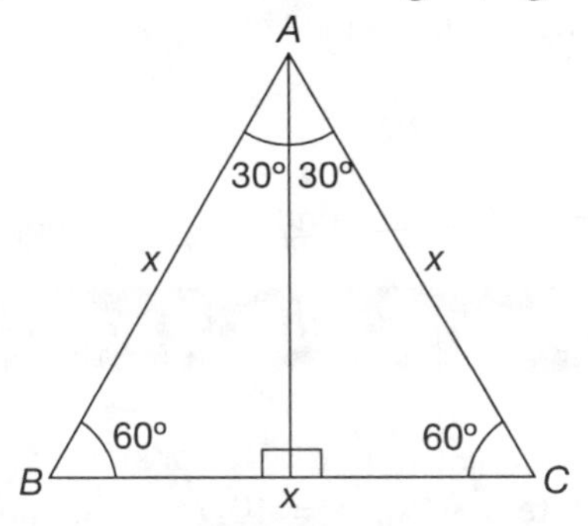

Figure 20.13

Theorem: The area of an equilateral triangle equals $\frac{\sqrt{3}}{3}$ times the square of the altitude of the triangle (Figure 20.14).

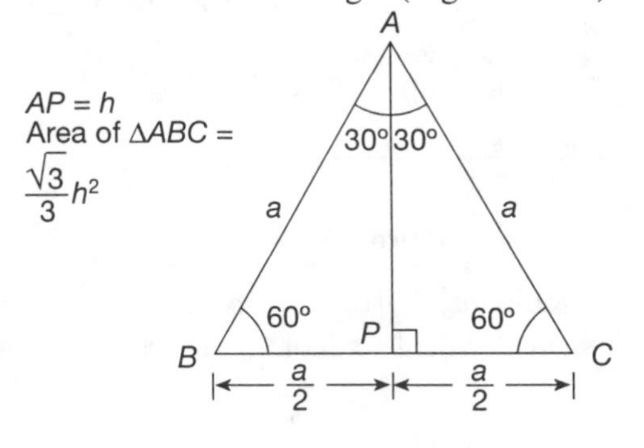

Figure 20.14

Theorem: A median drawn to a side of a triangle divides the triangle into two triangles of equal area (Figure 20.15).

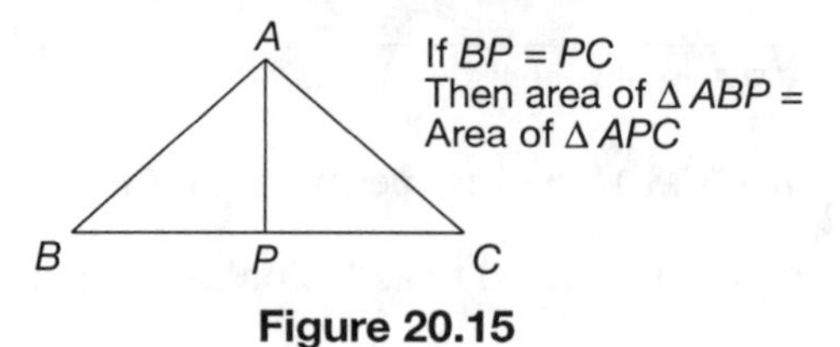

Figure 20.15

Theorem: The area of an isosceles triangle whose congruent sides have length l, with included angle α is:

$$A = \frac{1}{2} l^2 \sin \alpha.$$

Area is also given by the formula

$$A = h^2 \tan \frac{\alpha}{2},$$

where h is the length of the altitude to the side opposite angle α.

Problem Solving Examples:

Solve the oblique triangle ABC for side c and the two unknown angles, where $a = 20$, $b = 40$, $\alpha = 30$; and α is the angle between sides b and c.

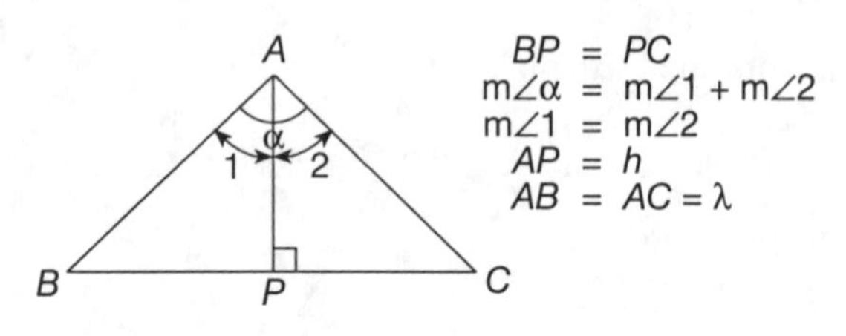

Figure 20.16

If we draw an altitude h from b to c, as in the accompanying diagram, we can find the length of this altitude by trigonometry.

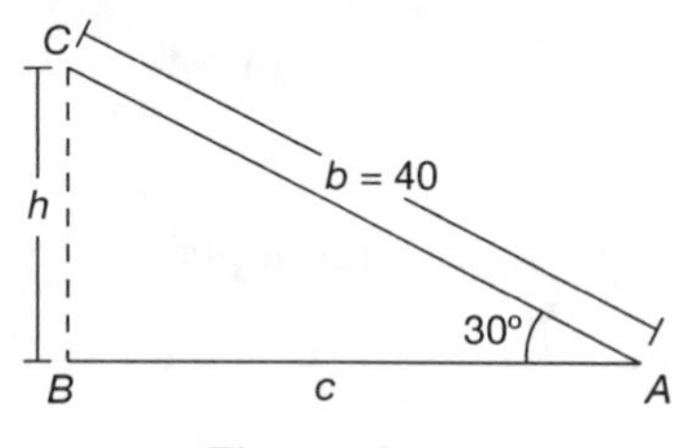

Figure 20.17

$$\sin 30° = \frac{h}{b}$$

$$h = b \sin 30°$$

Since $\sin 30° = \frac{1}{2}$, and $b = 40$,

$$h = 40\left(\tfrac{1}{2}\right) = 20 = \text{side } a.$$

Thus, the triangle must have the altitude as one of its sides; therefore, we have a right triangle with angles 30°, 60°, and 90°. Sides of such right triangles are in proportion $1 : \sqrt{3} : 2$, and the lengths are therefore $20 : 20\sqrt{3} : 40$. Hence, $c = 20\sqrt{3}$, and the two unknown angles are 60° and 90°.

Q Solve triangle ABC, given $a = 30$, $b = 50$, $\angle C = 25°$.

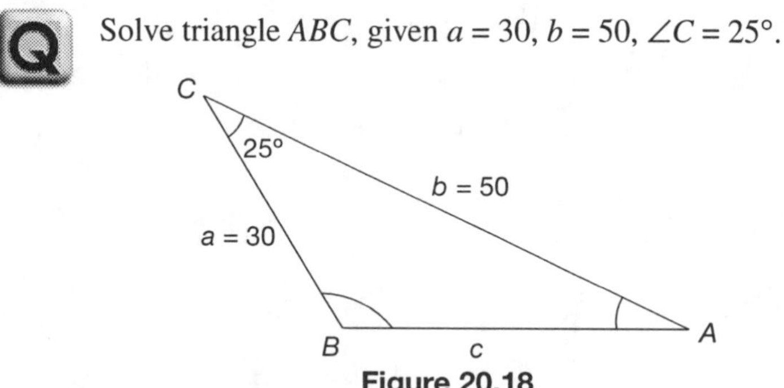

Figure 20.18

A Two of the sides of ΔABC and their included angle are given. We wish to find the third side, c. Therefore, use the law of cosines to find c.

$$c^2 = a^2 + b^2 - 2ab \cos C$$

$$c^2 = 30^2 + 50^2 - 2(30)(50) \cos 25°$$

$$c^2 = 900 + 2{,}500 - 2(30)(50)(0.9063)$$

$$c^2 = 681.1$$

$$c = 26 \text{ (to two significant digits)}$$

Use the law of sines to find one of the remaining angles.

$$\frac{\sin A}{30} = \frac{\sin 25^\circ}{26}$$

$$\sin A = \frac{30 \sin 25^\circ}{26} = \frac{30(0.4226)}{26}$$

$$\sin A = 0.4876$$

$$\angle A = 29^\circ \text{ (to the nearest degree)}$$

$\angle B$ can be found from $\angle A$ and $\angle C$.

$$\angle A + \angle B + \angle C = 180^\circ$$

$$\angle B = 180^\circ - \angle A - \angle C$$

$$= 180^\circ - (\angle A + \angle C)$$

$$= 180^\circ - (29^\circ + 25^\circ)$$

$$= 180^\circ - (54^\circ)$$

$$\angle B = 126^\circ$$

In triangle ABC, if $a = 675$, $\alpha = 48^\circ 36'$, find b, c, and β.

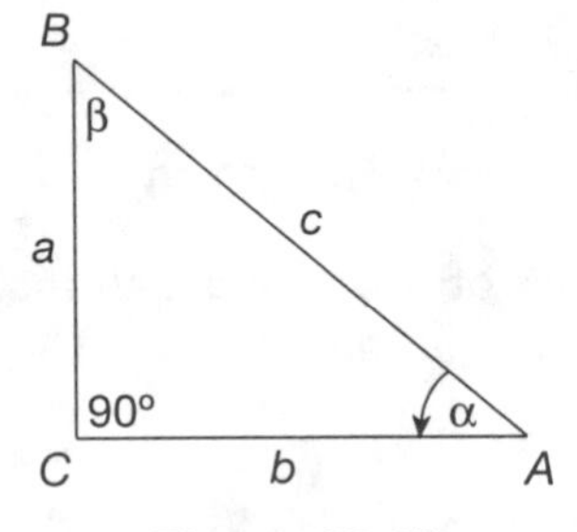

Figure 20.19

If we wish to work with α, the functions of α involving a and one other side are

$$\sin \alpha = \frac{\text{opposite}}{\text{hypotenuse}}$$

$\sin \alpha = a/c$

$$\tan \alpha = \frac{\text{opposite}}{\text{adjacent}}$$

$\tan \alpha = a/b$, and

$$\cot \alpha = \frac{1}{\tan \alpha} = \frac{\text{adjacent}}{\text{opposite}}.$$

$\cot \alpha = b/a$. Either of these ratios can be chosen. If the first is chosen, $\sin \alpha = a/c$ or $c \sin \alpha = a$ and $c = a/\sin \alpha$. Thus,

$$c = \frac{675}{\sin 48^\circ\ 36'} = \frac{675}{0.75011} = 899.85\ .$$

Using $\tan \alpha = a/b$ or $b = a/\tan \alpha$, we obtain

$$b = \frac{675}{\tan 48^\circ\ 36'} = 595.08\ .$$

To find β,

$$\begin{aligned} \alpha + \beta + 90^\circ &= 180^\circ \\ \alpha + \beta &= 180^\circ - 90^\circ = 90^\circ \\ \beta &= 90^\circ - \alpha = 90^\circ - 48^\circ 36' \\ \beta &= 41^\circ 24'\ . \end{aligned}$$

Quiz: Trigonometry and Trigonometric Functions-Solving Triangles

1. In the right triangle below, determine the measure of $\angle A$ and $\angle B$.

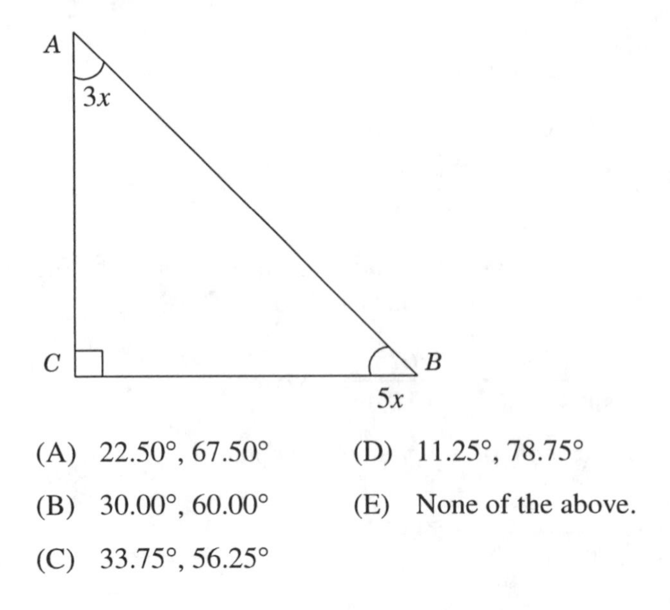

(A) 22.50°, 67.50°
(B) 30.00°, 60.00°
(C) 33.75°, 56.25°
(D) 11.25°, 78.75°
(E) None of the above.

2. In the triangle shown below, cos ω is equal to

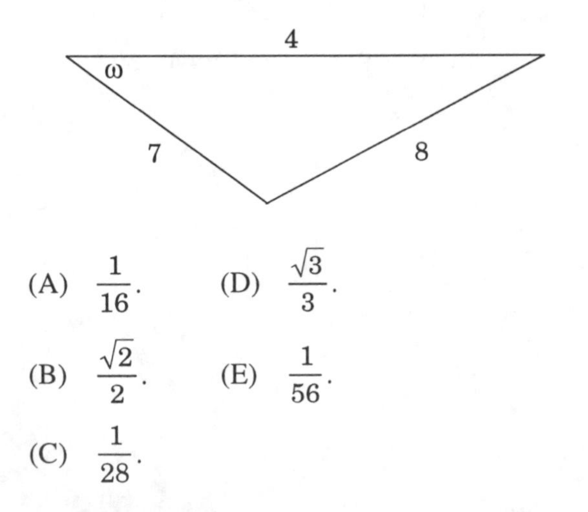

(A) $\frac{1}{16}$.
(B) $\frac{\sqrt{2}}{2}$.
(C) $\frac{1}{28}$.
(D) $\frac{\sqrt{3}}{3}$.
(E) $\frac{1}{56}$.

3. Given the triangle below, what is the value of θ?

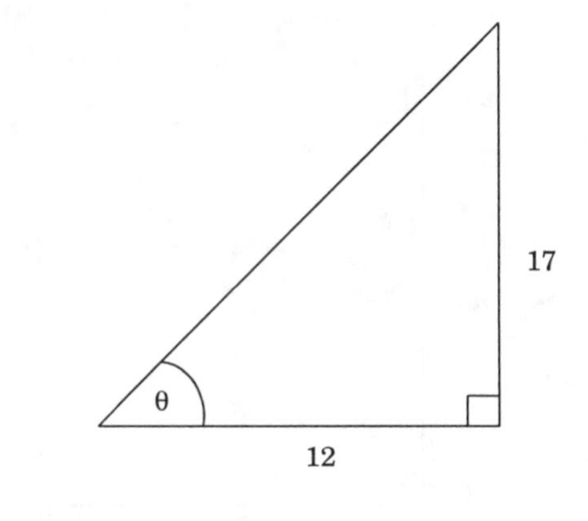

(A) 54.8°
(B) 35.2°
(C) 45.0°
(D) 42.8°
(E) 59.2°

4. In the figure, side y is 10 less than x. What is the length of x?

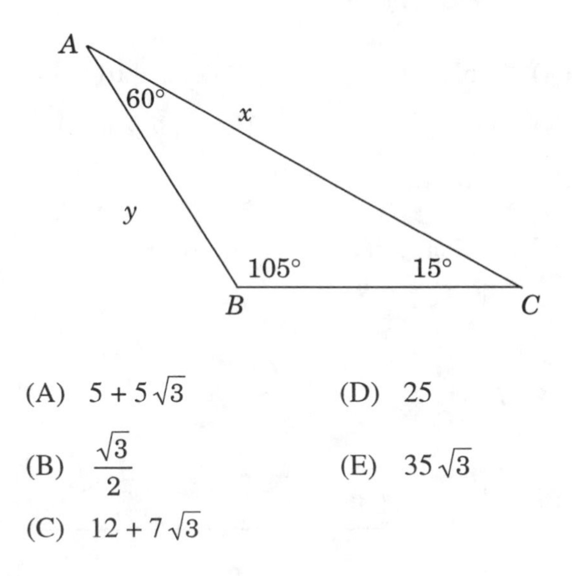

(A) $5 + 5\sqrt{3}$
(B) $\frac{\sqrt{3}}{2}$
(C) $12 + 7\sqrt{3}$
(D) 25
(E) $35\sqrt{3}$

5. In the right-angled triangle, c is equal to which of the following?

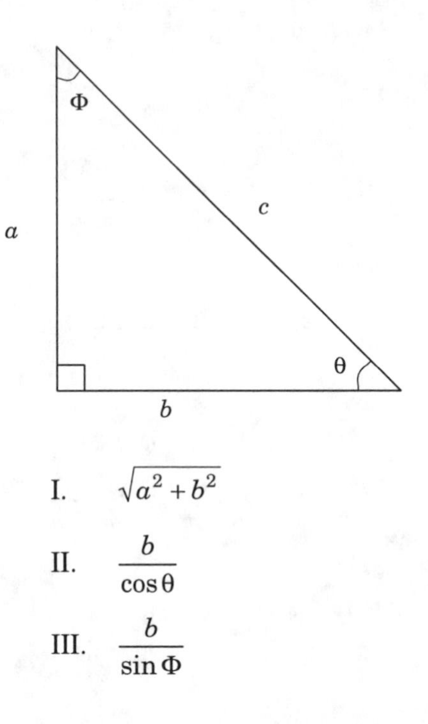

I. $\sqrt{a^2 + b^2}$

II. $\dfrac{b}{\cos\theta}$

III. $\dfrac{b}{\sin\Phi}$

(A) I only.
(B) II only.
(C) III only.
(D) I and III only.
(E) I, II, and III.

6. If $\alpha = 63.7°$, what is the length of a?

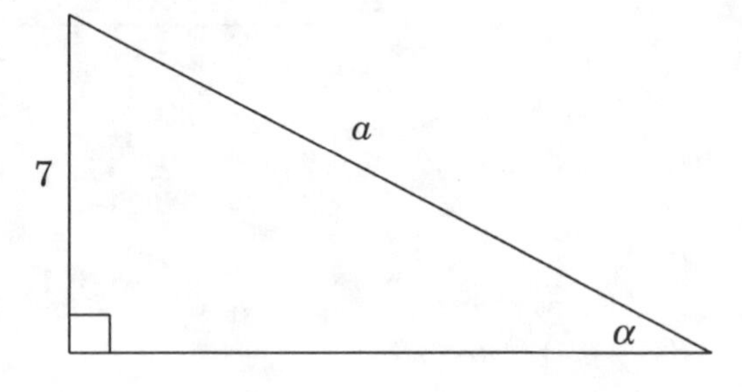

(A) 15.80
(B) 7.81
(C) 8.70
(D) 0.13
(E) 5.21

7. Evaluate: $\sqrt{1-\cos^2(60°)}$.

(A) 0.5
(B) –0.5
(C) 0.866
(D) –0.866
(E) Undefined

8. Evaluate 2cos(θ) sin(θ) when θ = 15.

(A) 1.9318
(B) 0.2588
(C) 0.5
(D) –0.2588
(E) –0.5

9. If $0° < \theta < 90°$ and $\sin\theta = \frac{a}{b}$, then $\tan\theta =$

(A) $\frac{a}{\sqrt{a^2-b^2}}$.
(B) $\frac{a}{\sqrt{b^2-a^2}}$.
(C) $\frac{a}{a^2+b^2}$.
(D) $\frac{b}{\sqrt{a^2-b^2}}$.
(E) $\frac{b}{\sqrt{b^2-a^2}}$.

10. If $0° \le \theta \le 45°$ and $\sin 2\theta = \frac{\sqrt{3}}{2}$, then $(\cos\theta + \sin\theta)^2 =$

(A) $1-\frac{\sqrt{3}}{2}$.
(B) 1.
(C) $1+\frac{\sqrt{3}}{2}$.
(D) $1+\sqrt{3}$.
(E) $1+2\sqrt{3}$.

ANSWER KEY

1. (C)
2. (E)
3. (A)
4. (A)
5. (E)
6. (B)
7. (C)
8. (C)
9. (B)
10. (C)

CHAPTER 21

Inverse Trigonometric Functions and Trigonometric Equations

21.1 Inverse Trigonometric Functions

By definition, if for every number y in the range of a function f there is exactly one number x in the domain of f such that $y = f(x)$, then f is one-to-one. Then f has an inverse function f^{-1} whose domain is the range of f and whose range is f's domain.

If the domain of the six trigonometric functions is not restricted, then they have no inverses, because they repeat and thus are not one-to-one. If, however, the domains are restricted to suitable intervals, we grant ourselves one-to-one correspondence, and we can define an inverse for the intervals.

Here are the trigonometric functions and their inverses:

If $y = \sin x$, then $x = \sin^{-1} y = \text{Arcsin } y$

$y = \cos x$, $x = \cos^{-1} y = \text{Arccos } y$

$y = \tan x$, $x = \tan^{-1} y = \text{Arctan } y$

The others are written the same way.

21.1.1 Definition of Inverse Trigonometric Functions

1. Sine^{-1}

   $T_1 = \{(y, x) | (x, y) \text{ satisfies } y = \sin x, -\frac{\pi}{2} \leq x \leq \frac{\pi}{2}\}$

2. Cosine^{-1}

   $T_2 = \{(y, x) | (x, y) \text{ satisfies } y = \cos x, 0 \leq x \leq \pi\}$

3. Tangent^{-1}

   $T_3 = \{(y, x) | (x, y) \text{ satisfies } y = \tan x, -\frac{\pi}{2} < x < \frac{\pi}{2}\}$

4. Cotangent^{-1}

   $T_4 = \{(y, x) | (x, y) \text{ satisfies } y = \cot x, 0 < x < \pi\}$

5. Secant^{-1}

   $T_5 = \{(y, x) | (x, y) \text{ satisfies } y = \sec x, \ 0 \leq x < \frac{\pi}{2}, \frac{\pi}{2} < x \leq \pi\}$

6. Cosecant^{-1}

   $T_6 = \{(y, x) | (x, y) \text{ satisfies } y = \csc x, \frac{-\pi}{2} \leq x < 0, 0 < x \leq \frac{\pi}{2}\}$

Capital letters at the beginning indicate the functions defined above.

The graph of the standard inverse trigonometric functions are shown by solid portions of the curves given in Figures 21.1. through 21.6.

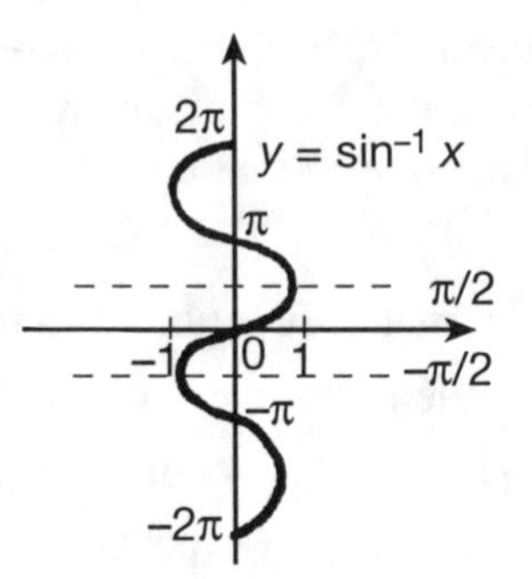

Figure 21.1

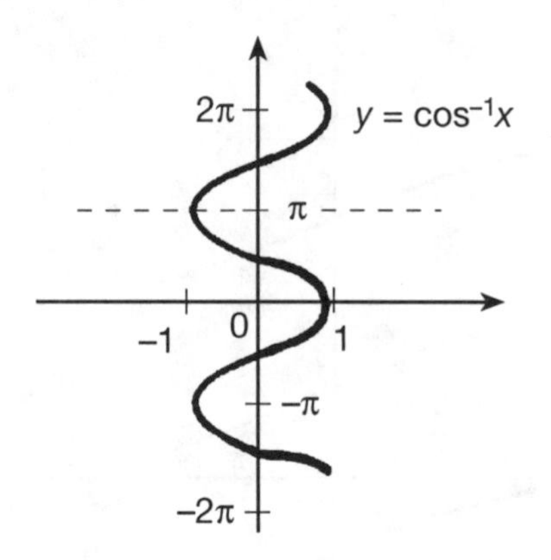

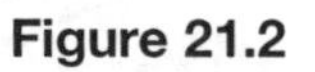

Figure 21.2

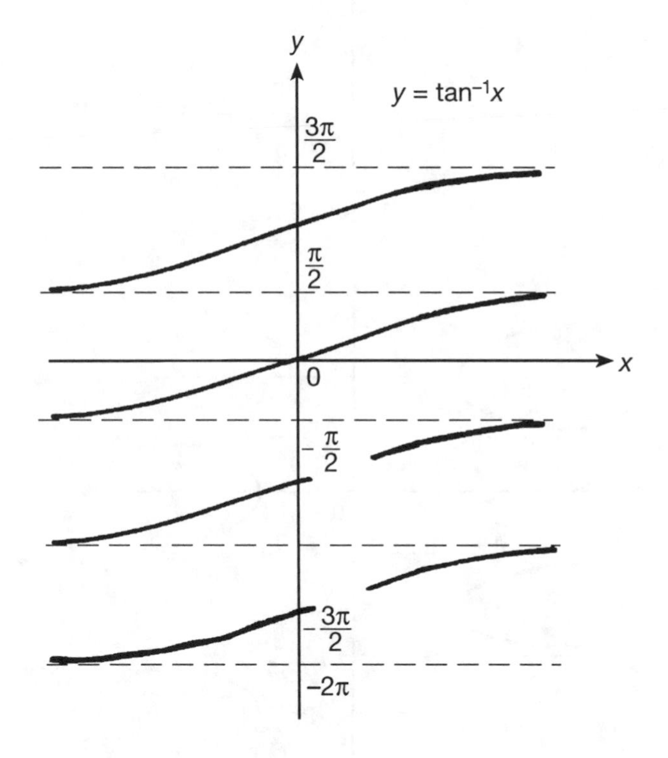

Figure 21.3

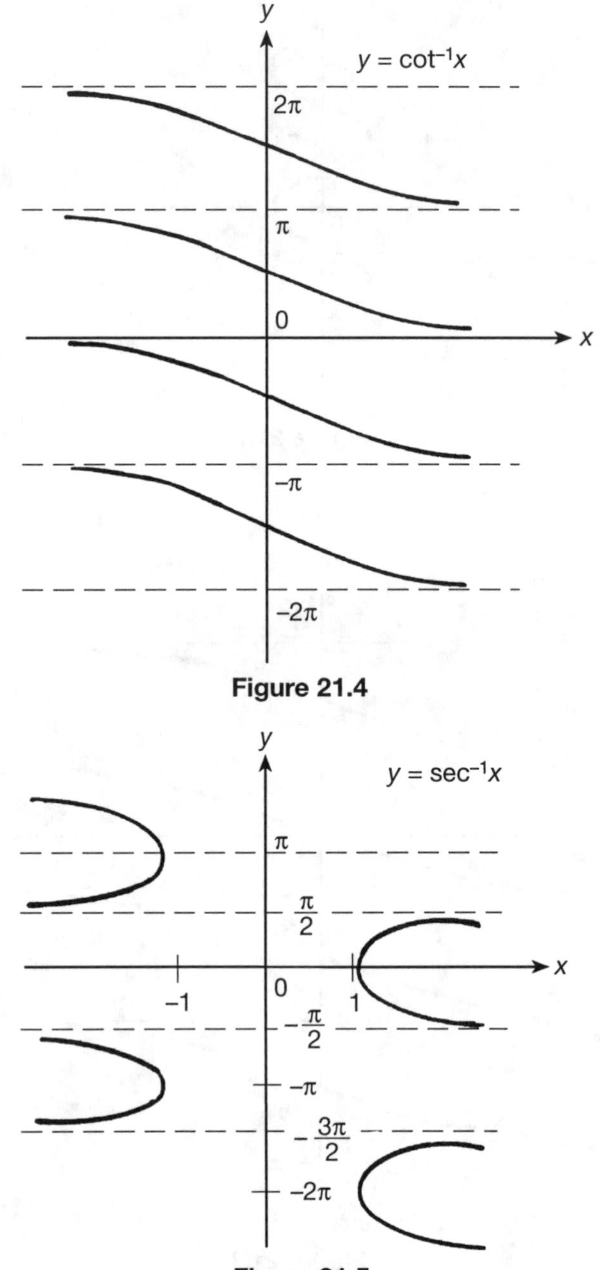

Figure 21.4

Figure 21.5

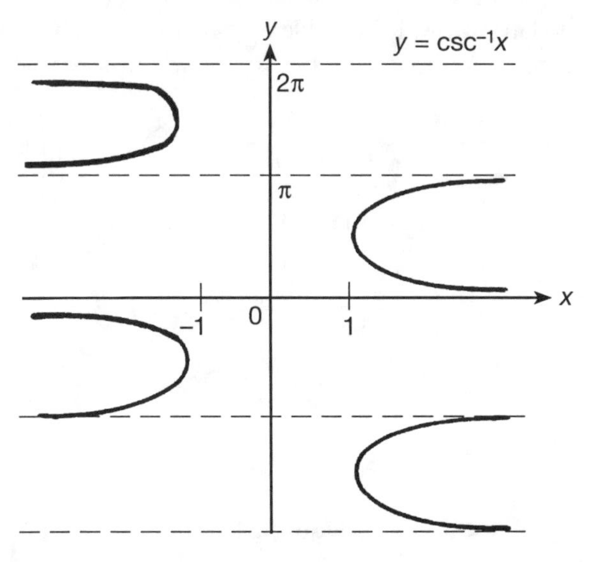

Figure 21.6

Problem Solving Examples:

Calculate the following numbers:

A) Arctan $\sqrt{3}$ C) Tan^{-1} 1.871

B) Tan^{-1} .2027

A) The expression $\tan y = x$ is equivalent to $\arctan x = \tan^{-1}x = y$. Let the expression $\arctan\sqrt{3} = y$. Hence, the expression $\arctan\sqrt{3} = y$ is equivalent to $\tan y = \sqrt{3} = 1.7321$. In a table of trigonometric functions, the number y that corresponds to $\tan y = 1.7321$ is approximately 1.05 radians.

B) The expression $\tan^{-1}.2027 = \arctan .2027$. Let the expression $\tan^{-1}.2027 = y$. Hence, the expression $\tan^{-1}.2027 = \arctan .2027 = y$ is equivalent to $\tan y = .2027$. In a table of trigonometric functions, the number y that corresponds to $\tan y = .2027$ is approximately .20 radians.

C) The expression $\tan^{-1}1.871 = \arctan 1.871$. Let the expression $\tan^{-1}1.871 = y$. Hence, the expression $\tan^{-1}1.871 = \arctan 1.871 = y$ is

equivalent to tan $y = 1.871$. In a table of trigonometric functions, the number y that corresponds to tan $y = 1.871$ is approximately 1.08 radians.

In ΔABC (shown in Figure 21.7), $A = \text{arc cos}\left(-\frac{\sqrt{3}}{2}\right)$. What is the value of A expressed in radians?

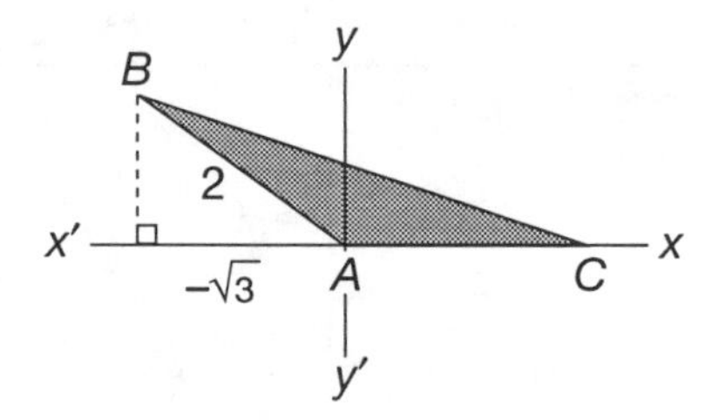

Figure 21.7

A The expression "arc cos $\left(-\frac{\sqrt{3}}{2}\right)$" means "the angle whose cosine equals $-\frac{\sqrt{3}}{2}$." Angles whose cosine equals $-\frac{\sqrt{3}}{2}$ are 150°, 210°, –150°, and –210°.

Since the principal value of an arc cosine of an angle is the positive angle having the smallest numerical value of the angle, 150°, or $\frac{5\pi}{6}$, is the principal value of angle A.

21.2 Trigonometric Equations

A trigonometric equation is an equation which involves one or more trigonometric functions of an unknown angle.

Trigonometric equations are of two types—the identity and the conditional equation.

An identity is a true statement for all values in the replacement set for which the equation is defined.

Example: $\sin^2\theta + \cos^2\theta = 1$

A conditional equation is a true statement only for particular values or sets of the variable quantities involved.

Example: $\sin\theta + \cos\theta = 1$

Example: Solve the equation

$$\sin^2\theta + 2\cos\theta - 1 = 0$$

for non-negative values of θ less than 2π.

Solution: Two trigonometric functions of the unknown θ itself appear in this equation. Accordingly, we make use of the identity connecting these functions, namely,

$$\sin^2\theta + \cos^2\theta = 1,$$

to transform it into an equation involving only one function of θ. Replace $\sin^2$ by $1 - \cos^2\theta$.

$$\sin^2\theta + 2\cos\theta - 1 = 0$$

$$1 - \cos^2\theta + 2\cos\theta - 1 = 0$$

Factor out $\cos\theta$:

$$\cos\theta\,(2 - \cos\theta) = 0.$$

Whenever a product of two numbers $ab = 0$, either $a = 0$ or $b = 0$; hence, $\cos\theta = 0$ or $2 - \cos\theta = 0$. Thus, $\cos\theta = 0$ or $\cos\theta = 2$.

Now there are two angles in the range $0 \le \theta < 2\pi$ for which $\cos\theta = 0$, namely,

$$\theta = \frac{\pi}{2}, \qquad \theta = \frac{3\pi}{2}.$$

But, since a cosine of an angle can never exceed unity, the relation $\cos\theta = 2$ does not yield a value of θ. Hence, we have just two solutions, as given above. It is easy to check these solutions.

Check: $\sin^2\theta + 2\cos\theta - 1 = 0$

$$(\sin\theta)^2 + 2\cos\theta - 1 = 0$$

For $\theta = \frac{\pi}{2}$

$$(\sin \frac{\pi}{2})^2 + 2 \cos \frac{\pi}{2} - 1 = 0$$

$$1 + 2 \times 0 - 1 = 0$$

$$0 = 0 \checkmark$$

For $\theta = \frac{3}{2}\pi$

$$(\sin \frac{3}{2}\pi)^2 + 2\cos \frac{3}{2}\pi - 1 = 0$$

$$(-1)^2 + 2 \times 0 - 1 = 0$$

$$0 = 0 \checkmark$$

Problem Solving Examples:

Show that if $x > 0$, then Arctan x = Arccot $\frac{1}{x}$.

A Let u = Arctan x. Then tan $u = x$, and $-\frac{1}{2}\pi < u < \frac{1}{2}\pi$. Since $x > 0$, $0 < u < \frac{1}{2}\pi$. Observe that

$$\frac{1}{x} = \frac{1}{\tan u} = \cot u,$$

and $0 < u < \frac{1}{2}\pi$. Since $\frac{1}{x} = \cot u$ and $0 < u < \frac{1}{2}\pi$ we have u = Arccot $\left(\frac{1}{x}\right)$. But we have already stated that u = Arctan x. Hence,

Arctan x = Arccot $\frac{1}{x}$.

Find the solution set on $(0, 2\pi)$ for $\sin x = \cos x$.

A Since the equation does not lend itself to factoring, we divide both sides by $\cos x$, obtaining $\sin x/\cos x = 1$, or $\tan x = 1$. The solution set of this new equation is $\{\pi/4, 5\pi/4\}$. However, in dividing the original equation by $\cos x$, it was assumed that $\cos x \neq 0$ because division by 0 is not permitted. However, $\cos x$ may be equal to zero. When $\cos x = 0$, $x = \pi/2$ or $x = 3\pi/2$. Checking the four solutions $x = \pi/4, \pi/2, 5\pi/4$, and $3\pi/2$ in the original equation:

$\sin \pi/4 = 0.7071 = \cos \pi/4$ ✓

$\sin \pi/2 = 1 \neq \cos \pi/2$ since $\cos \pi/2 = 0$

$\sin 5\pi/4 = -0.7071 = \cos 5\pi/4$ ✓

$\sin 3\pi/2 = -1 \neq \cos 3\pi/2$ since $\cos 3\pi/2 = 0$

Therefore, the solution set of the original equation is:

$(\pi/4, 5\pi/4)$.

CHAPTER 22
Introduction to Analytic Geometry

Analytic geometry refers to the study of geometric figures using algebraic principles. This approach, commonly referred to as coordinate geometry, has been accredited to a seventeenth-century French mathematician, René Descartes.

Postulate: The points on a straight line can be placed in a one-to-one correspondence with real numbers such that for every point of the line there corresponds a unique real number, and for every real number there corresponds a unique point of the line.

A number scale is a straight line on which distances from a point are numbered in equal units; positively in one direction and negatively in the opposite direction. The origin is the zero point from which all distances are measured.

The Cartesian product of a set X and a set Y is the set of all ordered pairs (x, y) where x belongs to X and y belongs to Y.

The graph of $\mathbb{R} \times \mathbb{R}$ is called the Cartesian coordinate plane (Figure 22.1). Graphically, it consists of a pair of perpendicular lines called coordinate axes, and the plane they lie in. The vertical axis is the y-axis and the horizontal axis is the x-axis. The point of intersection of these two axes is called the origin. It is the zero point of both axes. Furthermore, points to the right of the origin on the x-axis and above the origin on the y-axis represent positive real numbers. The negative numbers are represented similarly below and to the left of the origin. Each element of the set $\mathbb{R} \times \mathbb{R}$ is represented by a point on the Cartesian coordinate plane.

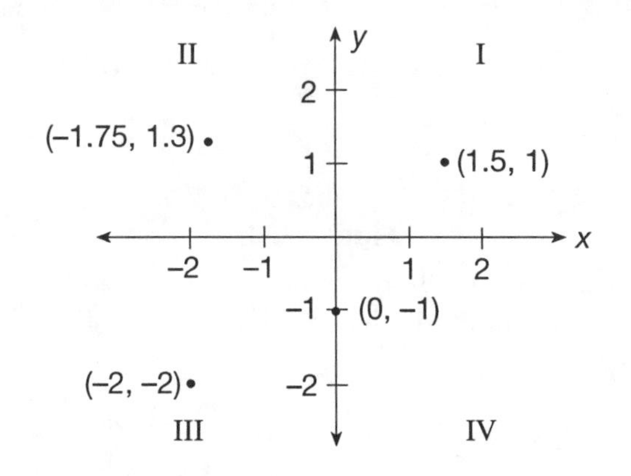

Figure 22.1

The abscissa is the x-coordinate of a given point, and the ordinate is the y-coordinate of the given point.

The plane is divided into four regions by the coordinate axes. These regions are numbered from one to four starting from the region where both x and y values are positive and proceeding counterclockwise.

The distance between any two points on a number scale is the absolute value of the difference between the corresponding numbers.

Theorem 1: For any two points A and B with coordinates (x_A, y_A) and (x_B, y_B) respectively, the distance between A and B is

$$d(A, B) = \sqrt{(x_A - x_B)^2 + (y_A - y_B)^2}\,.$$

This is commonly known as the distance formula (graphed in Figure 22.2).

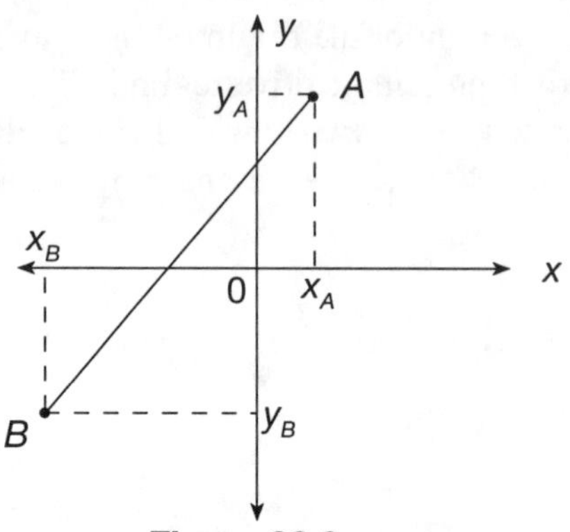

Figure 22.2

Theorem 2: Given a line segment with endpoints (x_A, y_A) and (x_B, y_B), the coordinates of the midpoint of the line segment are (x_m, y_m) where

$$x_m = \frac{x_A + x_B}{2}, \qquad y_m = \frac{y_A + y_B}{2}.$$

This is commonly known as the midpoint formula (graphed in Figure 22.3).

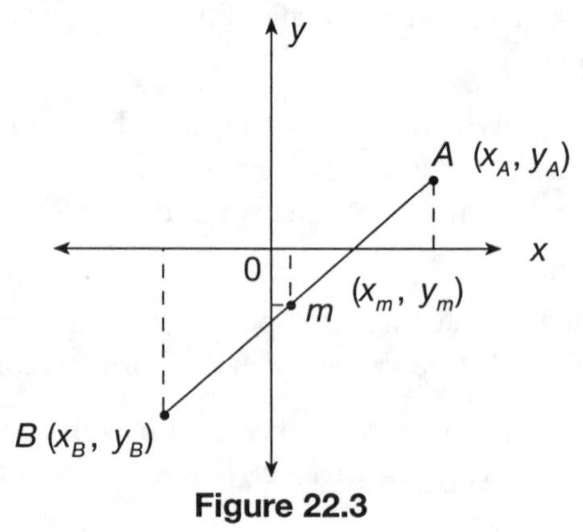

Figure 22.3

The distance d between a given point A with coordinates (x_A, y_A) and a line l defined as $ax + by + c = 0$ is given by the following formula:

$$d = \left| \frac{ax_A + by_A + c}{\sqrt{a^2 + b^2}} \right|$$

A parabola is the locus of points whose distance from a fixed line, called the directrix, and a fixed point, called the focus, is equal (Figure 22.4).

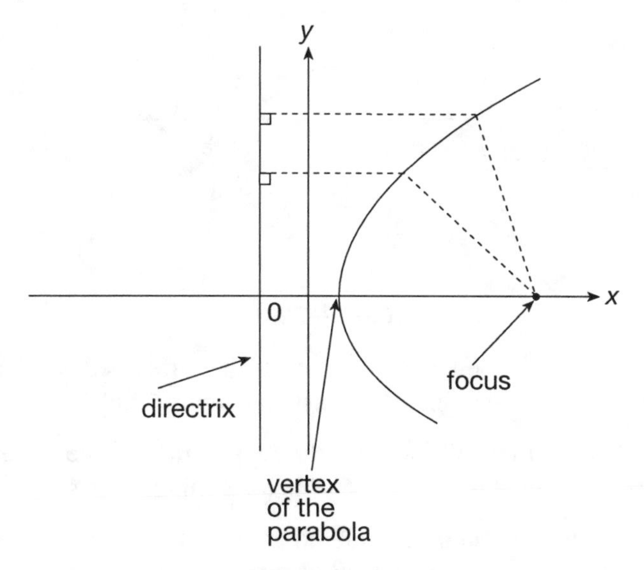

Figure 22.4

An ellipse is the locus of points, the sum of whose distances from two fixed points (foci) is a constant (Figure 22.5).

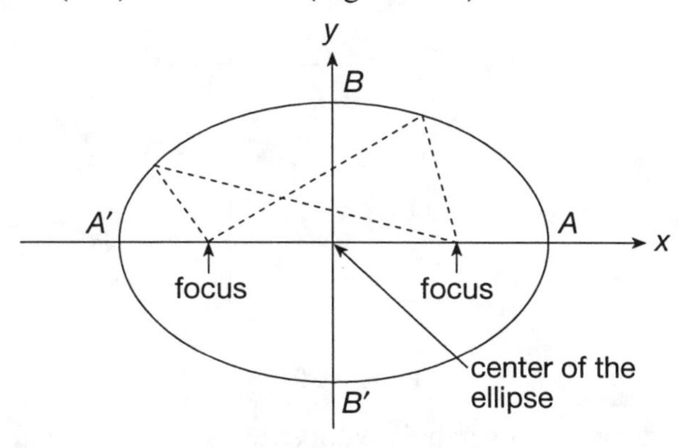

Figure 22.5

The line segment BB' is called the minor axis and the line segment AA' is called the major axis.

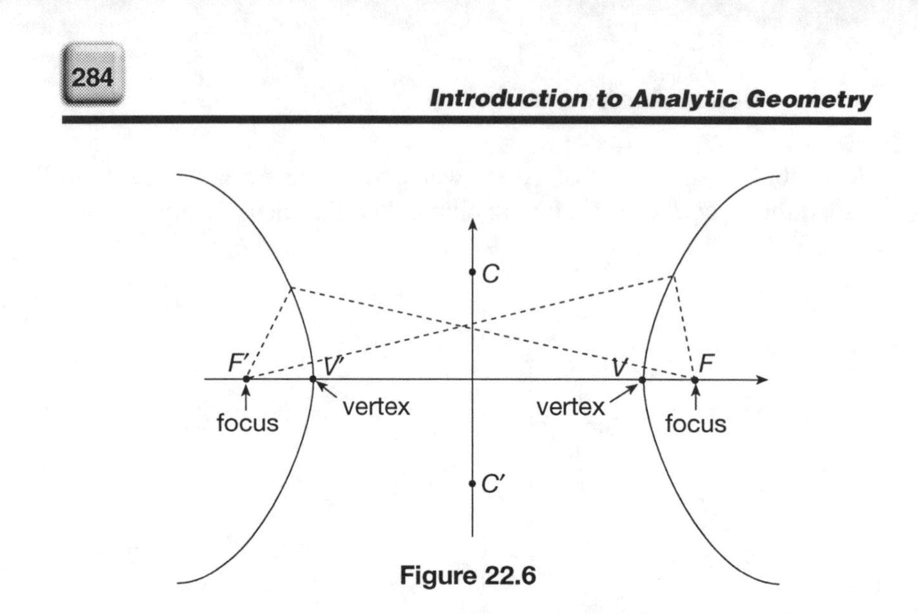

Figure 22.6

A hyperbola is the locus of all points the difference of whose distance from two fixed points is a constant (Figure 22.6).

The line segment VV' joining the two vertices is the hyperbola's transverse axis and the segment CC' is the conjugate axis.

The equation of a circle centered at (x_0, y_0) with radius r is given by: $(x - x_0)^2 + (y - y_0)^2 = r^2$, where x_0 and y_0 are the coordinates of the center of the circle and r is the length of the radius (Figure 22.7).

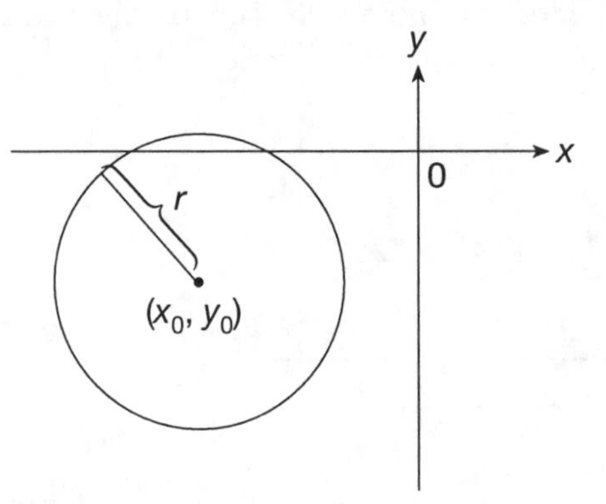

Figure 22.7

The equation of a parabola with vertex V at (x_0, y_0) and directrix d at $x = x_0 - p$ is

$$(y - y_0)^2 = 4P(x - x_0).$$

Note that the focus is then at $(x_0 + p, y_0)$ (Figures 22.8 and 22.9).

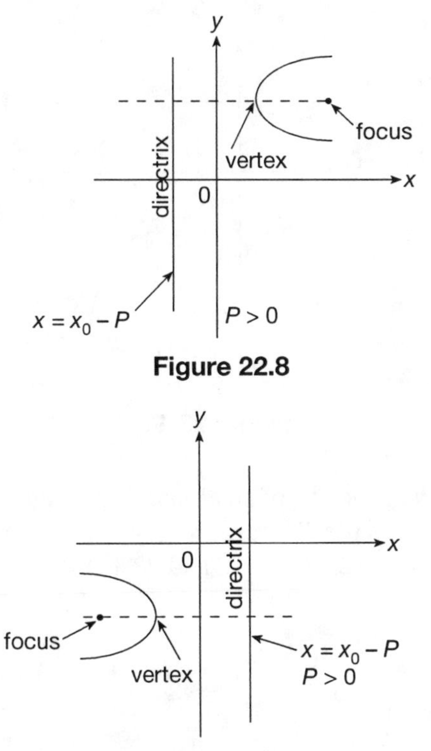

Figure 22.8

Figure 22.9

The equation of a parabola with vertex, $V(x_0, y_0)$ (Figure 22.10), and directrix, d, $y = y_0 - p$ (Figure 20.11) is

$$(x - x_0)^2 = 4P(y - y_0).$$

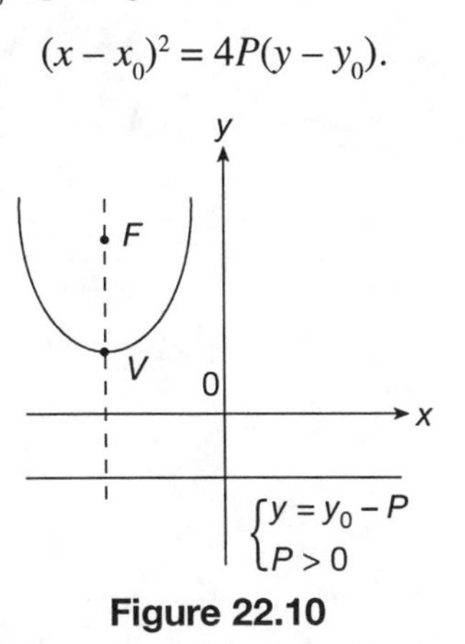

Figure 22.10

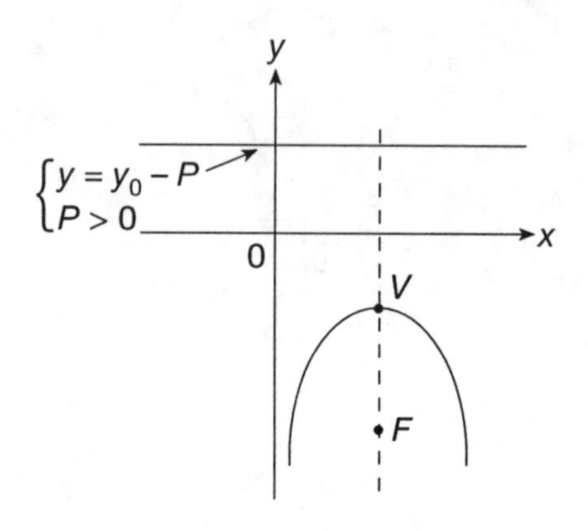

Figure 22.11

Theorem: The equation of an ellipse with center $C(x_0, y_0)$ and the major and minor axes parallel respectively to the x-axis and y-axis are given below (Figures 22.12, 22.13, and 22.14):

A) $\dfrac{(x-x_0)^2}{a^2}+\dfrac{(y-y_0)^2}{b^2} = 1, a \geq b > 0$

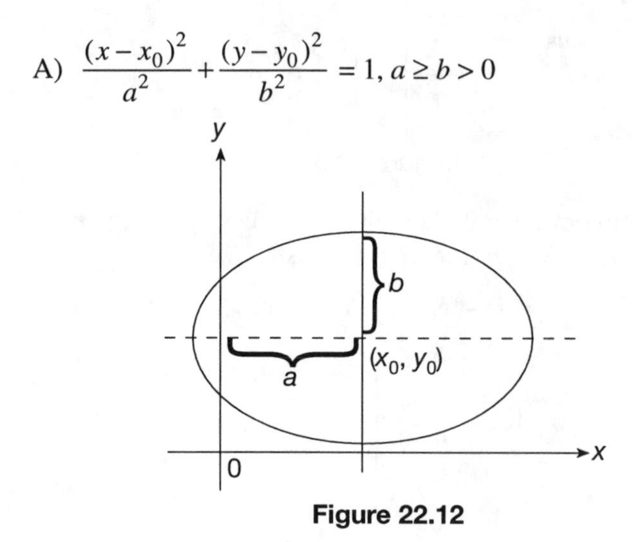

Figure 22.12

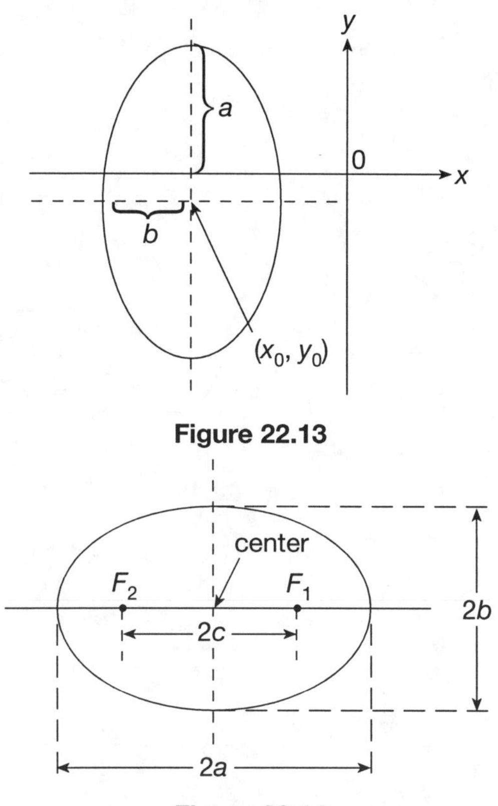

Figure 22.13

Figure 22.14

B) $\dfrac{(y-y_0)^2}{a^2}+\dfrac{(x-x_0)^2}{b^2}=1,\ a \geq b > 0$

where the length of the major axis is $2a$ and the length of the minor axis is $2b$. c = $\sqrt{a^2 - b^2}$ is half the distance between the two foci.

Theorem: The equation of a hyperbola with center $C(x_0, y_0)$, with transverse axis parallel to the x-axis is:

$$\frac{(x-x_0)^2}{a^2}-\frac{(y-y_0)^2}{b^2}=1$$

where the length of the transverse axis is $2a$ and the length of the conjugate axis is $2b$ (Figure 22.15).

$c = \sqrt{a^2 + b^2}$ is half the distance between the two foci F_1 and F_2 (Figure 22.16).

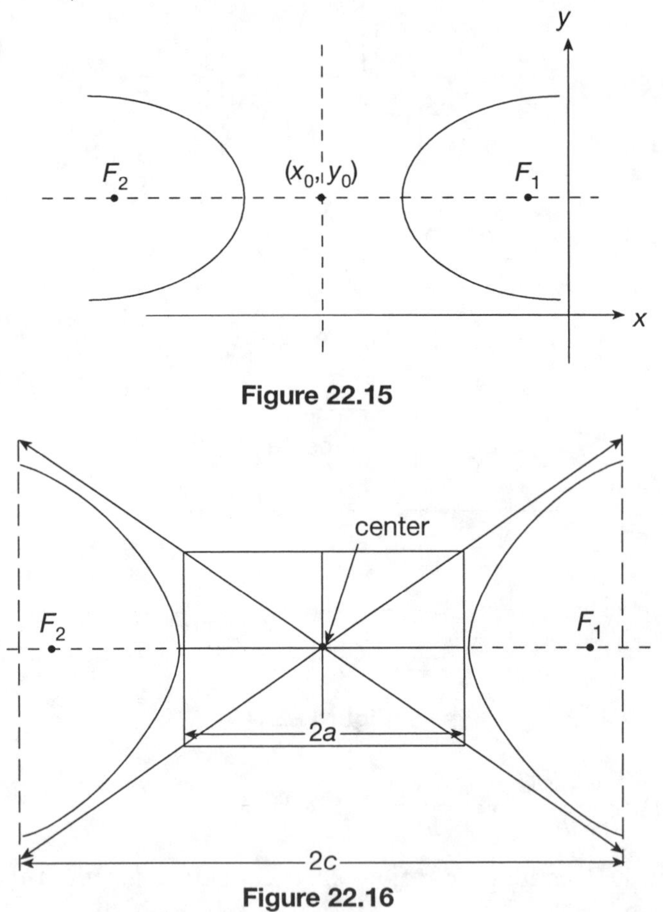

Figure 22.15

Figure 22.16

A point is expressed in polar form, (r, θ), as a distance and an angle instead of an x-coordinate and a y-coordinate. Draw a line from the point to the origin. r is the length of that line θ is the angle the line makes positively (counterclockwise) with the positive x-axis (Figure 22.17).

For example (5, 60°) represents the point five units from the origin and 60° around from the positive x-axis.

To change rectangular to polar coordinates:

$$r = \sqrt{x^2 + y^2}$$

$$\theta = \tan^{-1} \frac{y}{x}$$

To change polar to rectangular coordinates:

$$x = r \cos \theta$$

$$y = r \sin \theta$$

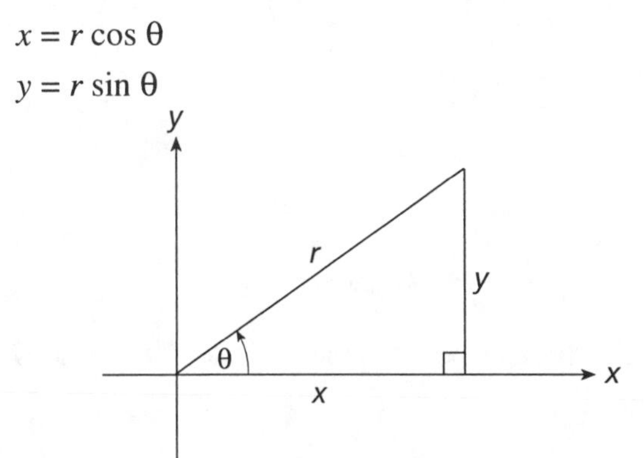

Figure 22.17

Problem Solving Examples:

A line segment AB is $7\frac{1}{2}$ in. long. Locate the point C between A and B so that AC is $\frac{3}{2}$ in. shorter than twice CB.

A See Figure 22.18.

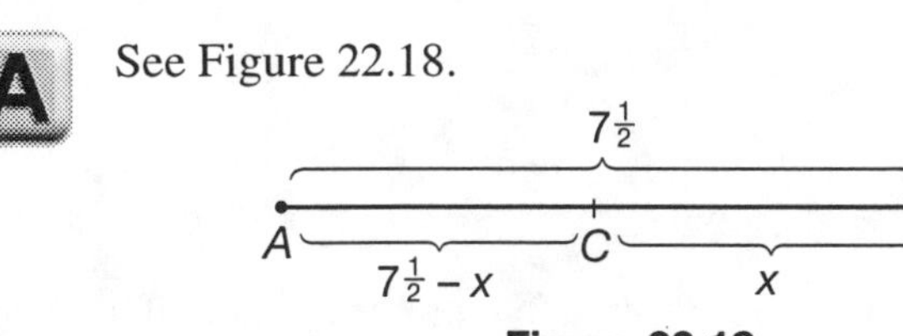

Figure 22.18

Let x = the length of CB in inches. Then $7\frac{1}{2} - x$ is the length of AC. We are told AC is $\frac{3}{2}$ in. shorter than twice CB. Thus, $AC = 2x - \frac{3}{2}$. Therefore:

$$7\frac{1}{2} - x = 2x - \frac{3}{2}$$

$$\frac{15}{2} - x = 2x - \frac{3}{2}.$$

Multiplying both members by 2,

$$15 - 2x = 4x - 3$$

$$-6x = -18$$

$$x = 3.$$

Therefore, $CB = 3$ and $AC = 7\frac{1}{2} - 3 = 4\frac{1}{2}$. Hence, C is located $4\frac{1}{2}$ inches from A and 3 in. from B.

Q What is the distance between the points $P(-4, 5)$ and $Q(1, -7)$?

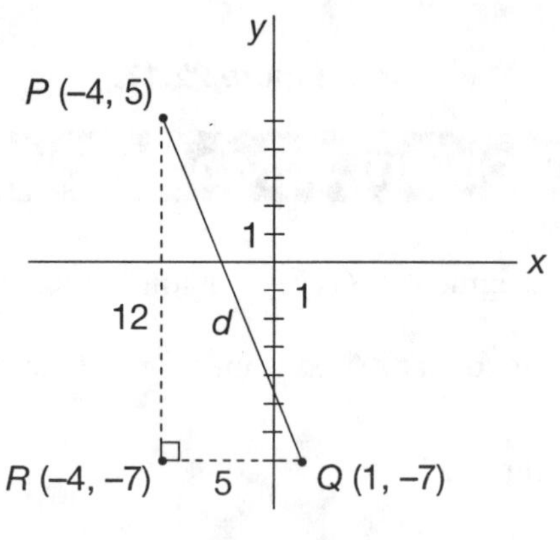

Figure 22.19

A Observe Figure 22.19. P being four units to the left of the Y-axis and Q being one unit to the right, the horizontal distance between P and Q is five units. Similarly, the vertical distance between P and Q is 12 units. The Pythagorean Theorem states that the sum of the squares of the legs of a right triangle equals the square of the hypotenuse. Thus, in right triangle PQR,

$$(\overline{PR})^2 + (\overline{RQ})^2 = (\overline{PQ})^2.$$

$$(\overline{PQ})^2 = (12)^2 + (5)^2 = 144 + 25 = 169$$

Taking the square root of both sides, $\overline{PQ} = 13$. Thus, the distance between (– 4, 5) and (1, –7), $\overline{PQ}$, is 13.

Quiz: Inverse Trigonometric Functions—Analytic Geometry

1. Perform the indicated operation: $\dfrac{x^2 - y^2}{2x} \times \dfrac{4x^2}{x+y}$.

   (A) $2x(4x^2 - y)$
   (B) $2x^2 + y$
   (C) $2x^2 - y$
   (D) $2x(x - y)$
   (E) $2x(x + y)$

2. Reduce $\dfrac{3x-6}{x^2-4}$.

   (A) $\dfrac{3}{x+2}$
   (B) $\dfrac{3}{x-2}$
   (C) $\dfrac{-3}{x+2}$
   (D) $\dfrac{-3}{x-2}$
   (E) Not given here.

3. Evaluate $\sin^{-1}\dfrac{\sqrt{3}}{2}$.

   (A) 0.866
   (B) 60°
   (C) 1.155
   (D) 0.6
   (E) –60°

4. Find $\sin^{-1} 0.4075$.

(A) 24° 3′
(B) –66° 57′
(C) 66° 57′
(D) –24° 3′
(E) Not given here.

5. What is the value of $\tan\left(\text{Arc}\cos\frac{\sqrt{2}}{2}\right)$?

(A) 45°
(B) .45
(C) .707
(D) 1
(E) – 0.707

6. Evaluate $\cos\left(\sin^{-1}\frac{1}{2}\right)$.

(A) .500
(B) –0.866
(C) $-\frac{\sqrt{3}}{2}$
(D) 60°
(E) $\frac{\sqrt{3}}{2}$

7. Find the midpoint of the segment from $R(-3, 5)$ to $S(2, -8)$.

(A) $\left(-\frac{1}{2},-\frac{3}{2}\right)$
(B) $\left(\frac{1}{2},-\frac{3}{2}\right)$
(C) $\left(\frac{1}{2},\frac{3}{2}\right)$
(D) $\left(-\frac{1}{2},\frac{3}{2}\right)$
(E) Not given here.

8. Find the distance between $P(5, 3)$ and $Q(8, 7)$.

(A) 25
(B) 5
(C) 4
(D) 3
(E) 16

9. Which of the following is an equation of a circle with radius r and with center at point $C\ (j, -k)$?

(A) $(x - j)^2 + (y - k)^2 = r$
(B) $(x - j)^2 + (y + k)^2 = r^2$
(C) $(x - j)^2 + (y + k)^2 = r^2$
(D) $(x + j)^2 + (y + k)^2 = r^2$
(E) $(x + j)^2 + (y - k)^2 = r^2$

10. Write the equation of a circle with center at $(-1, 3)$ and radius 9.

(A) $(x + 1)^2 + (y - 3)^2 = 81$
(B) $(x - 1)^2 + (y + 3)^2 = 9$
(C) $(x - 1)^2 + (y - 3)^2 = 81$
(D) $(x + 1)^2 + (y + 3)^2 = 81$
(E) $(x + 1)^2 + (y - 3)^2 = 9$

ANSWER KEY

1. (D)
2. (A)
3. (B)
4. (A)
5. (D)
6. (E)
7. (A)
8. (B)
9. (C)
10. (A)